『十二五』国家重点图书出版规划项目

中国古建筑测绘大系·宗教建筑

清华大学建筑学院 编写

廖慧农　王贵祥　刘畅 主编

五台山佛教建筑

国家出版基金项目

中国建筑工业出版社

Traditional Chinese Architecture Surveying and
Mapping Series:
Religious Architecture

MOUNT WUTAI'S BUDDHIST
ARCHITECTURE

Compiled by School of Architecture, Tsinghua University
Edited by LIAO Huinong , WANG Guixiang, LIU Chang

China Architecture & Building Press

「十二五」国家重点图书出版规划项目

Contents

目　录

Introduction

导言

Since its inception in 1946, the School of Architecture at Tsinghua University has been committed to surveying and mapping traditional Chinese buildings, following the practice of the Society for the Study of Chinese Architecture (*Zhongguo Yingzao Xueshe*) that LIANG Sicheng, a driving force in the Society and founder of Tsinghua's architecture department (known as the School of Architecture since 1988), and his assistant MO Zongjiang brought with them to Tsinghua. Between 1930 and 1945, with members of the Society, LIANG visited over two thousand Chinese sites located in more than two hundred counties and fifteen provinces, and discovered, identified and mapped over two hundred groups of traditional buildings, including the famous Tang-period east hall (dating to 857) of Foguang Monastery at Mount Wutai—which was not an easy task because of the harsh working conditions in the secluded and relatively inaccessible villages in the countryside. In that same spirit, despite the difficult political circumstances from the 1950s through the 1970s, the School of Architecture conducted a systematic survey of historical buildings in the New Summer Palace (Yiheyuan). At the beginning of the Cultural Revolution in the late 1970s, all members of the faculty focusing on the history of architecture went to Hebei province under the leadership of MO Zongjiang to measure and draw the main hall of Geyuan Monastery in Laiyuan, an important Liao-period relic hidden in the remote mountains. This was followed by in-depth research and analysis. At the same time, those professors that specialized in Chinese architectural history (MO Zongjiang, XU Bo'an, LOU Qingxi, ZHANG Jingxian, and GUO Daiheng) led a group of graduate students to Zhengding in Hebei province, where they conducted component analysis and research of Moni Hall at Longxing Monastery, a Northern-Song timber-frame structure that had partially collapsed but was then in the process of being rebuilt. They also investigated nearby

因为前辈学者梁思成及其助手莫宗江两位先生从中国营造学社继承的传统，清华大学建筑学院自创立以来，一直十分注重古代建筑实例的实地考察与测绘。尽管在 20 世纪 50 至 70 年代受到各种因素的影响与冲击，那时的清华大学建筑系，还坚持了对颐和园内一批古代建筑实例的系统测绘。改革开放刚刚开始的 1970 年代末，清华大学建筑历史方向的全体教师，就在莫宗江先生带领下，共同远赴偏僻的河北山区，考察测绘了创建于辽代的涞源阁院寺大殿，并对这座辽代木构建筑进行了系统研究。同是在那一时期，建筑历史教研室的莫宗江、徐伯安、楼庆西、张静娴、郭黛姮等教师，带领研究生赴河北正定，除了对正在落架重修的北宋木构大殿隆兴寺摩尼殿的大木构件进行现场分析研究外，还对正定及周边的古建筑进行了系统考察与调研。这种由老先生带队，

historical buildings in and around Zhengding. This practice of teamwork—senior researchers, instructors, and (graduate) students participating in the investigation and mapping of traditional Chinese architecture side by side—became an academic tradition at the School of Architecture of Tsinghua University.

Since the 1980s, fieldwork has been a crucial part of undergraduate education at the School, and focus and quality of teaching has constantly improved over the past decades. In the 1990s, professors like CHEN Zhihua and LOU Qingxi carried out surveying and mapping in advance of (re)construction or land development on sites all across China that were endangered. Since the turn of the twenty-first century, the two-fold approach— attaching equal importance to practice (fieldwork) and theory (teaching)—was widened and deepened. Sites were deliberately chosen to maximize educational outcome, resulting in a broader geographical scope and spectrum of building types. In addition to expanding on the idea of vernacular architecture, special attention was paid to local (government-sponsored) construction of palaces, tombs, and temples built in the official style (*guangshi*) or on a large scale (*dashi*), and to modern architecture dating to the period between 1840 and 1949. Students and staff have accumulated a lot of experience and created high-quality drawings through this fieldwork.

In retrospect, we have completed surveys of several hundred monuments and sites built in the official dynastic styles of the Song(Jin), Yuan, Ming and Qing all across the country. Fieldwork was always combined with teaching. Among the architecture surveyed are the (single- and multi-story) buildings in front and on the sides of the Hall of Supreme Harmony in the Forbidden City in Beijing; the architecture at Changling, the mausoleum of emperor Jiaqing located at the Western Qing tombs in Yi county, Hebei province; the monasteries on Mount Wutai, Shanxi province, including Xiantongsi, Tayuansi, Luobingsi, Pusading, Nanshansi (Youguosi), and Longquansi; Zhongyue Temple, Songyang Academy, and Shaolin Monastery in Dengfeng, Henan province; Xiyue Temple, Yuquan Court, and the Taoist architecture on the peaks of Mount Hua in Weinan, Shaanxi province; Chongan Monastery, Nanjixiang Monastery, Jade Emperor Temple (Yuhuangmiao) in Shizhang, and the temples of the Two Transcendents (Erxianmiao) in Xiaohuiling and Nanshentou, all situated in Lingchuan county of Shanxi province; and the upper and lower Guangsheng monasteries and the Water God's Temple in Hongdong, Shanxi province. In recent years, we have developed a specialized interest in the study of religious architecture of Shanxi province and investigated almost a dozen privately- or government-sponsored Song and Jin sites

002

的学术传统。

1980 年代以来，清华大学建筑学院始终在本科教学环节中，坚持讲授古代建筑测绘这门经典课程。这一传统在 21 世纪初的这十几年中始终延续。如果说，20 世纪 90 年代由陈志华、楼庆西等教授带领的测绘教学，将相当的注意力放在了分布于全国多个省、市、自治区大量传统乡土村落建筑的抢救性测绘教学上，进入 21 世纪以来，清华大学建筑学院开展的这种结合本科教学的古建筑测绘教学与实践，覆盖的地域范围与建筑类型范围更为宽广：除了进一步拓展乡土建筑的测绘之外，在对各地留存的历代官式或大式建筑，如宫殿、陵寝、寺庙等建筑的测绘以及近代建筑的测绘上，也积累了大量测绘经验、图纸及丰富的调研资料。以古代官式建筑测绘为例，结合本科教学，我们先后完成了北京故宫太和殿前及两侧门殿、楼阁与朝房建筑，河北易县清西陵昌陵完整建筑群，山西五台山显通寺、塔院寺、罗睺寺、菩萨顶、南山寺（佑国寺）、龙泉寺等多座整组寺院建筑群，河南登封中岳庙、嵩阳书院、少林寺古建筑群，陕西渭南华山西岳庙、玉泉院及华山山顶各道观古建筑群，山西陵川崇安寺、南吉祥寺、小会岭二仙庙、南神头二仙庙、石掌玉皇庙，以及山西洪洞广胜上寺、广胜下寺、水神庙等数百座古建筑实例的测绘，其时代的范围覆盖了宋（金）、元、明、清等历代木构建筑遗存实例。近几年，我们又将测绘的重点放在了高平、晋城等晋中及晋东南地区，

教师与研究生集体参与、对古代建筑进行深入考察与测绘研究的做法，在清华大学形成了一个良好

located in central Shanxi (Jinzhong) and southeastern Shanxi (Jindongnan), specifically in Gaoping and Jincheng counties. This includes the Youxian, Chongming, and Kaihua monasteries and the Two Transcendents Temple in Xilimen. Additionally, supported by the State Administration of Cultural Relics, the head of the Architecture History and Historic Preservation Research Institute at the School of Architecture, Liu Chang, led a group of students to map and draw the main hall of Zhenguo Monastery in Pingyao, a rare example from the Five Dynasties period. The survey results have been published. Tsinghua fieldwork in Shanxi has become an annual event that is jointly organized almost every summer by the faculty of the School of Architecture, including professors engaged in research on non-Chinese architecture, in cooperation with their graduate students.

It is worth mentioning that since 2007, the School has worked in collaboration with the well-known company China Resources Snow Breweries Ltd., which supports the transmission and dissemination of knowledge on traditional Chinese architecture and provides funds for the School's research and field investigation activities. Drawing on the support from industry allowed us greater initiative and flexibility, and we were thus able to carry out research on and survey often overlooked but no less important Song-Jin monuments in central and southeastern Shanxi.

Our years-long fieldwork has not only enabled us to teach students subject knowledge about scale, material, form, and decoration of traditional Chinese architecture as well as a sense of appreciation for the old, but has also provided us with plenty of data for monument preservation practice and research. China Architecture and Building Press spared no effort in compiling and publishing the results of the fieldwork in 2012. Publication has also been supported by the National Publishing Fund. This highlights not only the importance of our contribution to architectural education at the national level but also shows its significance for the transmission, development, and revival of traditional Chinese architectural culture both at home and abroad. In order to expand the reach of this work to an international audience, *the Traditional Chinese Architecture Surveying and Mapping Series* is being published bilingually. Based on the past ten years of fieldwork, we have now compiled five volumes, namely *Mount Wutai's Buddhist Architecture* (Traditional architecture on Mount Wutai, Shanxi), *Architecture Complex of Songshan* (Traditional architecture in Dengfeng, Henan), *Mount Hua's Yuemiao and Taoist Temples* (Traditional architecture on Mount Hua, Shaanxi), *Architecture Complex of Hongtong* (Traditional architecture in Hongtong, Shanxi), and *Architecture Complex*

对包括高平游仙寺、崇明寺、开化寺、西李门二仙庙等在内的十余座宋金建筑群，进行了全面而系统的测绘。这一期间，在国家文物局的支持下，建筑历史与文物保护研究所刘畅老师还带领研究生对五代时期创建的平遥镇国寺大殿等建筑进行了精细测绘，并出版了测绘研究成果。此外，清华大学建筑学院的测绘工作，几乎每年都是由全体建筑历史教师共同合作，并带领研究生们共同完成的。从事外国建筑史教学的老师，也不例外。

特别值得一提的是，自2007年以来，清华大学建筑学院与国家知名企业华润雪花啤酒（中国）有限公司建立了良好的合作关系。该集团不仅支持中国古建筑知识的传承与普及工作，也对清华大学建筑学院中国古代建筑研究及古建筑测绘工作给予了直接的支持，使得我们的古建筑测绘工作变得更为主动和更具选择性。一大批珍贵的山西晋中及晋东南地区宋金时代建筑实例的测绘与研究，就是在这样一个前提下得以顺利开展与完成的。

坚持数十年的古建筑测绘工作，不仅在培养学生对传统中国建筑的尺度、材料、造型与细部装饰的认知与感觉上起到了直接的影响，而且也为各地文物建筑保护与研究工作，提供了相当充分的资料支持。

2012年，中国建筑工业出版社花大气力组织了汇集全国重点院校建筑系古建筑测绘成果的中国古代建筑测绘大系的编辑出版工作。这一工作也获得了国家出版基金的支持。这不仅是对高校建筑教育成果的一份支持，也是对中国传统建筑文化传承、发展与复兴的一份支持。正是在这样一个背景与前提下，我们对近十余年来考察测绘的古代建筑案例加以整理，分别编汇了包括《五台山佛教建筑》《嵩山建筑群》《华山岳庙与道观》《洪洞建筑群》《高平建筑群》5册古建筑测

of Gaoping (Traditional architecture in Gaoping, Shanxi). The architectural drawings presented in these books are carefully selected and screened by Tsinghua professors. They only show a part of our comprehensive surveying and mapping work, but still cover a whole spectrum of geographic regions and time periods. Thus, they contain information of high academic value that may serve as a reference for future study and for the protection of cultural heritage. It is hoped that our work will help to promote interest in and improve understanding of traditional Chinese architecture, not only among Tsinghua students (through hands-on experiences in the fieldwork) but also among architectural historians and professionals engaged in monument preservation at home and abroad.

As a final thought, let me shortly address the workflow. The drawings presented here are based on survey and working sketches drawn up on site during several years of fieldwork conducted by Tsinghua professors together with graduate and undergraduate students. Back home, the measured drawings were redrawn over months of diligent work by graduate students with computer-aided software to achieve dimensionally accurate and visually appealing results, a project that was completed under the supervision of LIU Chang, head of Tsinghua's Architecture History Institute, and the Tsinghua professors LIAO Huinong and WANG Nan, as well as TANG Henglu and his colleagues from the WANG Guixiang Studio. We would like to take this opportunity to thank the professors, students and colleagues who participated in the fieldwork and its revision.

Our final thanks go to LI Jing, assistant researcher at the Architecture History Institute here at Tsinghua. Next to participating in surveying and mapping, she organized the development of the book and moreover, made this book possible in the first place.

WANG Guixiang, LIU Chang, LIAO Huinong
Architecture History and Historic Preservation Research Institute, School of Architecture,
Tsinghua University
December 5, 2017

Translated by Alexandra Harrer

绘图集，作为这套『中国古建筑测绘大系』的部分成果。尽管这只是我们多年测绘成果的一部分，

但也是清华建筑历史学科教师们仔细筛选、认真校对、充分整理之后的较具典型性与参考性的成果。

这些成果不仅地域覆盖面大，而且建筑遗存的时代跨度也相当长，具有十分重要的学术价值。希望

这些成果对高校建筑系学生们学习古建筑，建筑历史学者研究古建筑，以及文物保护工作者从事文

物古建筑的保护与修缮，能够起到积极的推动作用与重要的参考价值。

最后要提到的一点是，除了参与测绘的教师、研究生与本科生多年历尽辛苦的测量与绘图工作

之外，此次清华大学建筑学院承担的这 5 册测绘图集，也经由建筑历史与文物保护研究所刘畅、廖

慧农、王南和他们的研究生，以及王贵祥工作室团队的唐恒鲁等同仁们在既有测绘图纸基础上，经

过数月认真仔细的线条分层、图面调整、数据校对、图面完善等缜密修复工作，在这里也要向参加

测绘图整理的老师、同学和同事们表示感谢。

还应该特别提到的是建筑学院建筑历史与文物保护研究所的助理研究员李菁博士，她不仅参加

了多次测绘，还为这套书最后的编辑与出版做了大量相关工作。这里一并表示感谢。

清华大学建筑学院 建筑历史与文物保护研究所

王贵祥、刘畅、廖慧农

2017 年 12 月 5 日

Preface

Wutaishan or Mount Wutai (Mountain of the Five Terraces), historically known as Mount Qingliang (Mountain of the Cool and Cold), is situated in Wutai county, Xinzhou prefecture, Shanxi province, and part of the Taihang Mountains. Changes in the geological structure formed five flat-topped peaks roughly corresponding to the directions after which they are named— an east, west, south, and north peak surrounding a central peak. Mount Wutai is one of the Four Sacred Mountains in Chinese Buddhism. Among all the mountains in Chinese Buddhist history, it is the most famous one and often referred to as the first among the renowned Buddhist mountains. Mount Wutai has a long history. Early on in the Tang dynasty, it was recognized as the home of the bodhisattva of wisdom, Mañjuśrī, better known by his Chinese name Wenshu. For more than a thousand years, since the Tang and Song dynasties, Buddhism has flourished at Mount Wutai and Buddhist temples were built continuously and in large numbers.

In late imperial China, Mount Wutai was the subject of imperial pilgrimage and patronage, especially by the Qing emperors Kangxi and Qianlong, which endowed the monasteries on Mount Wutai with the fame and prestige they enjoy today. In spite of repair on several occasions in modern times, the cultural landscape on Mount Wutai still preserved the original Tang layout as a complex consisting of sixty-eight Buddhist monasteries. Moreover, its palatial-style halls (*diantang*), pagodas (*ta*), stone pillars inscribed with Buddhist scriptures (*jingchuang*), carvings and wall paintings have exceptional historic-cultural significance and artistic value. In 2006 (announced in the sixth batch), the State Administration of Cultural Relics designated the historical architecture of Mount Wutai, including Xiantong Monastery, as a National Priority Protected Site.

Religious architecture on Mount Wutai is concentrated in the small valley town of Taihuai, where all the large-scale temples are located such as the Xiantong, Tayuan, Guangzong, Luobing, and Zhuxiang monasteries and Pusading (Bodhisattva Summit) (Fig.1, Fig.2). Drawings of four architectural ensembles (Xiantongsi, Tayuansi, Luobingsi, and Bodhisattva Summit) are included in this volume of the Traditional Chinese Architecture Surveying and Mapping Series, and these Buddhist sites are introduced concisely below.

序言

五台山古称清凉山，位于山西省忻州市五台县境内，属太行山的支脉。由于历代地质结构的变化，形成五峰环列的高山，也就是现在的东台、西台、南台、北台和中台。五台山是中国四大佛教名山之一，在众多的佛教名山之中，它的香火最盛，知名度最高，可谓佛教名山之首。五台山历史悠久，早在唐代就被公认为是文殊菩萨的道场。唐宋以来的一千多年里，五台山的佛教活动一直兴盛繁荣，不断兴建寺院庙宇，成为一处佛寺云集的佛教名山。尤其清朝以后更是受到康熙和乾隆皇帝的重视和扶持，使得五台山的声望更加显赫，规模更加宏大。虽然近代以来五台山的佛寺也遭到了屡毁屡建的经历，但至今仍保存着唐代以来的68座佛寺，其中的殿堂、佛塔、经幢以及雕塑、壁画均具有极高的历史价值、文化价值和艺术价值。2006年国家文物局在发布第六批全国重点文物保护单位名单时，将五台山古建筑群与显通寺合并为同一个全国重点文物保护单位。五台山内的台怀镇为寺院集中之地，显通寺、塔院寺、广宗寺、罗睺寺、菩萨顶、殊像寺等重要的大型寺院都坐落于此（图1、图2）。本册展示的是显

图2　五台山景观　图片来源：敖仕恒摄于2010年6月

图1　台怀镇俯瞰（局部）　图片来源：李路珂提供

Fig.1 Overlooking the Taihuai Town (Partial) Source: Provided by LI Luke
Fig.2 Landscape of Mount Wutai Source: Photographed by AO Shiheng in June 2010

1. Xiantongsi or Monastery of Manifested Penetration

Xiantongsi (Monastery of Manifested Penetration) is the largest extant Buddhist complex on Mount Wutai. First built in 68 during the Yongping reign of emperor Ming of Han, then known as Dafu Lingjiusi (or Monastery of Great Faith at Vulture Peak), it was expanded under emperor Xiaowen of Wei and again renovated under the first Ming emperor who bestowed it with a plaque inscribed with the name Da Xiantongsi (Great Xiantong Monastery). After several name changes, the monastery became known under this name since 1687 (the twenty-sixth year of the Qing emperor Kangxi). It is one of the oldest Buddhist monasteries in China. In 1982, Xiantong Monastery was designated as a National Priority Protected Site by the State Administration of Cultural Heritage.

The complex comprises an area of about 80.000 m sq. The architecture adds up to more than four hundred bay-lengths. Most buildings date to the Ming-Qing period. The layout (of the palatial-style halls [diantang] and wing-rooms [xiangfang]) is strict and focuses on axial symmetry. Seven main buildings are aligned along the central axis—Guanyin (Avalokiteśvara) Hall, Wenshu (Mañjuśrī) Hall, Treasure Hall of the Great Hero (Daxiong baodian), a beamless hall (wuliangdian), Hall of Mañjuśrī with a Thousand Alms Bowls (Qianbo Wenshudian), followed by a golden (copper) hall and a sutra repository (tongdian and cangjingdian). East- and west-side halls (peidian) of lesser importance stand on the left and right sides. Daxiong baodian was rebuilt in 1899 (the twenty-fifth reign year of Qing emperor Guangxu) and is the largest of its kind on Mount Wutai. The golden hall was cast in 1610 (the thirty-eighth reign year of Ming emperor Wanli) and used 10.000 jin of copper. It is one of the best-preserved copper halls in China. Additionally, the largest copper bell and the largest incense burner of all the monasteries on Mount Wutai are housed here.

In more detail, the first building aligned along the main axis is Avalokiteśvara Hall, a five-bay wide structure with a gabled roof. A statue of the bodhisattva Avalokiteśvara (Guanyin) is placed in the center at a position corresponding to the central bay, and flanked by statues of Mañjuśrī (Wenshu) and Samantabhadra (Puxian) on the sides. Next is the five-bay, hip-gable roofed Mañjuśrī Hall that has a small projecting portico (baosha) attached to the rear of the hall. Inside are installed seven statues of the bodhisattva Mañjuśrī, including the Dazhi (Great Wisdom) Wenshu in the center, the Ganluo (Heavenly Nectar) Wenshu behind it, the Dharma Protector Skanda (Weituo) in front of it, and the Eighteen Arhats on both sides. Daxiong baodian is the third important building aligned along the

通寺、塔院寺、罗睺寺和菩萨顶这四座寺庙的测绘图。以下就清华大学建筑学院测绘的这四座寺院作一简要介绍。

一、显通寺

显通寺是五台山现存的最大寺院，始建于汉明帝永平年间（公元68年），初名大孚灵鹫寺，魏孝文帝时期扩建，明太祖重修，又赐额『大显通寺』。清康熙二十六年（1687年），改名为大显通寺，是中国最早的佛寺之一。显通寺早在1982年就被国家文物局公布为全国重点文物保护单位。

显通寺占地约8万平方米，各种建筑400余间，且大多为明、清时期的建筑。殿堂、厢房布局严整，中轴线分明，配殿左右对称，中轴线上的建筑依次为观音殿、文殊殿、大雄宝殿、无量殿、千钵文殊殿、铜殿和藏经殿等7座殿宇。其中大雄宝殿重建于光绪二十五年（1899年），是五台山最大的一座大雄宝殿。铜殿铸于明万历三十八年（1610年），共用铜10万斤，是中国国内保存最好的铜殿之一。寺中现存五台山最大的铜钟和最大的铜香炉。显通寺的第一座殿是观音殿，为五开间硬山。殿内中间供着观音菩萨像，左右供着文殊和普贤菩萨像。文殊殿是显通寺的第二座大殿，为五开间歇山，后带抱厦。文殊殿内供着七尊文殊菩萨像，其中正中为大智文殊，大智文殊后面是甘露文殊。文殊像前有护法神韦驮像，两侧有十八罗汉像。

大雄宝殿是显通寺的第三重大殿，为举行佛事活动的主要场所。大殿共七开间带周围廊，前有抱厦，重檐庑殿顶。面宽34.4米，进深21米。殿内佛坛上并列供奉着三尊佛像：中是释迦牟尼佛，西是阿弥陀佛，东是药师佛；两旁有十八罗汉像，背后有观音、文殊、普贤三尊菩萨像（图3）。

图3　显通寺大雄宝殿　图片来源：敖仕恒提供

Fig.3 Daxiong baodian of Xiantongsi Source: Provided by AO Shiheng

central axis and the main venue for the religious Buddhist activities in the monastery. The seven-bay structure is surrounded by a corridor with a projecting portico (*baosha*) in front, and crowned by a double-eaves hipped roof. It is 34.4 m wide and 21 m deep. Three Buddha statues sit on the altar inside the hall—Shakyamuni Buddha in the center, Amitābha Buddha to the west, and the Medicine Buddha or Bhaiṣajyaguru (Yaoshi) to the eas. On both sides stand the Eighteen Arhats, and behind them, the three bodhisattvas Avalokiteśvara, Mañjuśrī, and Samantabhadra (Fig.3).

The next building, the beamless hall, is a two-story, seven-bay domed brick structure covered with a double-eaves hip-gable roof. It is 28.8 m wide, 16 m deep, and 20.3 m tall. Inside the hall is enshrined the "boundless" Amitabha Buddha. Behind the beamless hall is the Hall of Manjushri with a Thousand Alms Bowls (known in Chinese as Qibo Wenshudian), the golden hall, and the sutra repository (Fig.4~Fig.6). The golden hall is a square double-eaves structure of 4.64 m width, 4.2 m depth, and about 8 m height. It is built in a unique style and a rare example of its kind in China. Because numerous small Buddha statues are carved on the interior walls of the hall, it is also known as Ten-thousand Buddha Hall.

Xiantong Monastery is the oldest religious complex on Mount Wutai. It is the place where Buddhist culture on Mount Wutai and in North China in general emerged and developed. Its eventful history tells the story of the origin and development of Buddhism on both the local and the interregional-state levels. Xiantong Monastery thus has significant historical, scientific and artistic value.

2. Bodhisattva Summit or Pusading

Bodhisattva Summit or Pusading, also known as Court or Cloister of the Bodhisattva's True Countenance, was first established in the reign of emperor Xiaowen of Northern Wei but was repaired and rebuilt on several occasions during the Tang, Song, Ming, and Qing dynasties. Located on top of Vulture Peak north of Xiantong Monastery in the town of Taihuai, it is the largest Lamaist monastery of Mount Wutai. A walkway of one hundred and eight steps leads from the foot of the mountain to the summit of Vulture Peak. A four-column archway (*paifang*) stands at the beginning of the walkway, and on the summit further along the central axis are situated a second archway, a front gate (*shanmen*), Hall of Heavenly Kings (Tianwangdian), Daxiong baodian, and Great Mañjuśrī Hall (Da Wenshudian). The architecture comprises one hundred and twenty-one bays.

二、菩萨顶

菩萨顶始建于北魏孝文帝时期，唐、宋、明、清各代都予以修葺或重建。菩萨顶又名真容院，整座建筑群坐落在台怀镇显通寺北侧的灵鹫峰之顶，是五台山最大的喇嘛寺院。灵鹫峰山脚下建有一条笔直的108级大石台阶蹬道直达峰顶。登上台阶蹬道迎面便是一座四柱牌坊。在菩萨顶中轴线上的建筑依次是牌坊、山门、天王殿、大雄宝殿和大文殊殿。全寺共有殿堂房屋121间。菩萨顶的大雄宝殿面阔三间，前出抱厦，并带有周围廊，面宽约15米，进深约16米，单檐歇山顶，殿顶覆盖琉璃黄瓦，殿内供释迦牟尼、弥陀、药师佛三佛（图7）。大文殊殿面阔三间，带周围廊，面宽约18米，进深约12.6米，单檐庑殿顶，

五台山显通寺无量殿分上、下两层，明七间，暗三间，重檐歇山顶，砖券结构。面宽28.8米，进深16米，高20.3米，殿内供有无量佛。显通寺后半部依次是千钵文殊殿、铜殿和藏经殿（图4~图6）。其中铜殿是比较有特点的建筑。铜殿平面为方形，双层檐。铜殿高约8米，宽4.64米，深4.2米。因殿内四壁铸满了佛像，也称万佛殿。显通寺是五台山佛教建筑群中最早的一座寺院，是五台山地区乃至中国北方广大地区佛教文化的发祥地和发展中心。显通寺的建立与发展是五台山佛教文化的缩影，显通寺也是研究中国北方佛教文化的活标本，它见证了佛教在中国发展的历程，具有较高的历史、科学和艺术价值。

图 5　显通寺铜殿及藏经殿　图片来源：贺从容提供

图 4　显通寺文殊殿　图片来源：廖慧农提供

图 7　菩萨顶大雄宝殿　图片来源：敖仕恒提供

图 6　显通寺藏经殿　图片来源：廖慧农提供

Fig.4　Wenshudian of Xiantongsi
　　　Source: Provided by LIAO Huinong
Fig.5　*Tongdian* and *cangjingdian* of Xiantongsi
　　　Source: Provided by HE Congrong
Fig.6　*Cangjingdian* of Xiantongsi
　　　Source: Provided by LIAO Huinong
Fig.7　Daxiong baodian of Bodhisattva Summit
　　　Provided by AO Shiheng

Daxiong baodian is a three-bay structure with a projecting front portico (*baosha*) and a surrounding corridor. It is almost square, about 15 m wide and 16 m deep, with single-eave hip-gable roof covered with glazed yellow tiles. Three Buddhas— Shakyamuni, Amitabha, and Bhaiṣajyaguru—sit inside the hall (Fig.7). Next, Great Mañjuśrī Hall is a three-bay structure surrounded by a corridor and measures 18 m in width and 12.6 m in depth. It has a single-eave hipped roof covered with glazed yellow tiles. The ridge is decorated with a golden shining dharma wheel made of copper. A statue of the bodhisattva Mañjuśrī riding a lion stands on top of the Buddhist dais inside the hall; the statue is made of clay and painted. The Eighteen Arhats flank the dais on the east and west sides. Great Mañjuśrī Hall is one of the places Buddhist worshippers must visit when on Mount Wutai. The monastery is the most important site of the Gelug School, or Yellow Hat sect, of Tibetan Buddhism on Mount Wutai. The buildings from the Kangxi period are of high rank and show the historical significance of Pusading as a destination of imperial pilgrimage.

3. Luohuosi or Monastery of Rahula, Son of Siddhartha Gautama

Luohousi (Monastery of Rahula, son of Siddhartha/son of the Buddha), first established in the Tang dynasty, was repaired and rebuilt on several occasions in succeeding periods. After entering through the temple gate, the buildings are aligned one after another along the central axis—a screen wall, Hall of Heavenly Kings (Tianwangdian), Wenshu (Mañjuśrī) Hall, Daxiong baodian, and a sutra repository *cangjingge*. The architectural structures measure up to a total of one hundred and eighteen bay-lengths.

The first building on the central axis is the Hall of Heavenly Kings, named after the four statues enshrined therein. The Heavenly Kings are also known as the Four Heavenly Guardians because they watch over the cardinal directions and protect the entrance to a monastery (Fig.8). The second building, Mañjuśrī Hall, houses a statue of the bodhisattva Mañjuśrī (Wenshu) and is a three-bay structure with surrounding corridor that is 16.8 m wide and 12.7 m deep (Fig.9). The third building is Daxiong baodian that is a three-bay structure with front gallery and square in ground plan, measuring 14.6 by 14.3 m. The Buddhas of the Past, Present and Future—Shakyamuni, Bhaiṣajyaguru, and Amitābha—are installed inside. The last building is the five-bay, two-story, gable-roofed sutra repository that contains Buddhist scriptures. Four Buddha statues sit on a lotus dais installed in the center of the building. When a mechanical mechanism is turned on,

---

殿顶覆盖琉璃黄瓦，殿脊正中置金碧辉煌的鎏金铜法轮，殿内佛坛上供文殊菩萨骑狻猊彩色泥塑像，东西两侧佛坛上则供十八罗汉塑像。菩萨顶的文殊菩萨殿，是朝山信徒必定要参诣礼拜之处。该寺是五台山黄庙中的首庙，其主要建筑的规格在康熙年间即为较高等级，可见其在皇家道场中的重要地位。

三、罗睺寺

罗睺寺始建于唐代，宋、明、清代都予以修葺或重建。进入罗睺寺山门后，在中轴线上的建筑有影壁、天王殿、文殊殿、大雄宝殿和大藏经阁，全寺共有殿堂房舍118间。寺内第一座殿为天王殿，内塑四大天王，也称四大金刚（图8）。第二座殿为文殊殿，内供文殊菩萨。文殊殿为三开间，有周围廊，面宽为16.8米，进深为12.7米（图9）。第三座为大雄宝殿，内供释迦牟尼佛、药师佛、阿弥陀佛，合称『三世佛』。大殿为三开间，带前廊，面宽为14.6米，进深为14.3米。第四座殿为大藏经阁，为五开间硬山二层阁楼。殿内正中高竖一朵莲花，内含四尊佛像，有时八瓣莲花会缓缓绽开，现出四方阿弥陀佛，呈现出『开花现佛』的神秘色彩。该寺为五大禅处之一，康熙时由青庙改为黄庙，是研究五台山藏传佛教文化的重要实例。

图9 罗睺寺文殊殿 图片来源：贺从容提供

图8 罗睺寺天王殿 图片来源：贺从容提供

Fig.8 Tianwangdian of Luohousi Source: Provided by HE Congrong
Fig.9 Wenshudian of Luohousi Source: Provided by HE Congrong

the lotus appears to bloom—the eight flower petals open slowly—and one can see four golden Amitābhas seated back-to-back. The spectacular arrangement represents the idea of Buddha manifesting himself in blossoming flowers (*kaihua xianfo*).

Originally one of the Five Great Chan Monasteries of Han Buddhism ("Blue-green Religion"), Luohuo Monastery became affiliated with Tibetan Buddhism ("Yellow Religion")—precisely the Gelug School or Yellow Hat sect—in the reign of the Qing emperor Kangxi. It is an important example of Tibetan Buddhist culture on Mount Wutai.

4. Tayuansi or Pagoda Court Monastery

Formerly a part of Xiantongsi, Tayuansi (Pagoda Court Monastery) is located to the south of it today. It became an independent monastic unit after renovation during the reign of the Ming emperor Wanli, and since then, has been known under its present name. The architecture comprises more than one hundred and thirty bay-lengths and covers an area of 15,000 m sq. The main buildings are aligned along a central axis and comprise a screen wall, a monumental arch, a front hall, Hall of Heavenly Kings (Tianwangdian), Treasure Hall of Great Compassion and Longevity (Daciyanshou baodian), and a sutra repository. Nearby are a bell tower and a drum tower, the Building of Mountains and Seas (or Shanhailou), and Mañjuśrī (Wenshu) Pagoda located in the east-side courtyard. The principal building of Tayuan Monastery is, as the name indicates, the Great White Pagoda (built in the Wanli reign period of the Ming dynasty) that stands in the main courtyard surrounded by corridor buildings (Fig.10~Fig.13). It is axially placed between the Treasure Hall of Great Compassion and Longevity and the sutra repository. The pagoda is also known as Great Compassion and Longevity Treasure Pagoda (Daciyanshou baota) and as Sakyamuni Buddha Relic Pagoda, named after the sacred relics (*sheli; sarira* [Sankrit]) it is supposed to safeguard. Shaped like an inverted bowl similar in form to a Lamaist stupa, the pagoda sits on a square, 31.4m wide pedestal. With a height of 54 m, the Great White Pagoda is regarded as a local landmark.

In 2005, the Ministry of Construction and the State Administration of Cultural Heritage applied for dual nomination of Mount Wutai on the World Cultural Heritage List and the World Natural Heritage List. As part of the preparation for nomination, the School of Architecture at Tsinghua University, under the leadership of WANG Guixiang, organized a team of professors (HE Congrong, LIU Chang) and graduate students (LI

四、塔院寺

塔院寺位于显通寺的南面，原为显通寺的一部分，明代万历年间重修时从显通寺独立出来，更名塔院寺。寺内中轴线上的主要建筑依次有影壁、牌坊、前殿、天王殿、大慈延寿宝殿、藏经阁。周围有钟鼓楼、山海楼及文殊寺塔等建筑。寺内主要建筑，大慈延寿宝殿在前，藏经阁在后，大白塔位居其中，全名叫释迦文佛真身舍利宝塔，又叫大慈延寿宝塔，为明代万历年间修建。大白塔状如覆钵，为喇嘛塔的造型，塔的基座为方形，边长 31.4 米，塔高约 54 米，被视为五台山的标志性建筑物。

2005 年建设部及国家文物局拟将山西五台山作为世界文化、自然双重遗产提名地向联合国教科文组织提交报告。清华大学建筑学院为配合此项工作，于 2005 年 7 月由王贵祥教授带队，组织师生前往五台山进行古建筑测绘。其中参与的教师还有贺从容、刘畅；研究生有李路珂、胡介中、姜东成、白颖、史韶华、郑亮。承担主要测绘任务的学生为 2002 级本科生，共 50 人，测绘了三组重要的古建筑群——显通寺、塔院寺及罗睺寺，完成图纸 128 张。这些图纸大部分用于由建设部及国家文物局组织、清华大学建筑学院景观学系杨锐教授主持编制的《五台山申遗文本》之中，为 2009 年 6 月五台山成功登录世界遗产名录起到了重要的支持作用。

周设廊屋，布局完整（图 10～图 13）。全寺共有殿堂楼房 130 余间，占地面积 15000 平方米。大白塔，

图二　塔院寺天王殿　图片来源：辛惠园提供

图 10　塔院寺牌楼　图片来源：李路珂提供

Fig.10　Archway of Tayuansi　Source: Provided by LI Luke
Fig.11　Tianwangdian of Tayuansi　Source: Provided by XIN Huiyuan

图13　塔院寺大白塔近景　图片来源：测绘学生提供

图12　塔院寺大白塔远景　图片来源：敖仕恒提供

Fig.12　The distant view of the Great White Pagoda (baita) of Tayuansi
　　　　Source: Provided by AO Shiheng
Fig.13　The close view of the Great White Pagoda of Tayuansi
　　　　Source: Provided by the surveying students

图14　菩萨顶工作照　图片来源：贺从容提供

Fig.14 Work photos of Bodhisattva Summit Source: Provided by HE Congrong

Luke, HU Jiezhong, JIANG Dongcheng, BAI Ying, SHI Shaohua, and ZHENG Liang) to survey and measure the historical buildings on Mount Wutai in July 2005. More than fifty bachelor students (Class of 2002) participated in the fieldwork that focused on three sites—Xiantongsi, Tayuansi, and Luohousi—and drew one hundred and twenty-eight architectural drawings. Most of them were published in *Wutaishan shenyi wenben* compiled by YANG Rui, professor of landscape design at the School of Architecture of Tsinghua University, in cooperation with the Ministry of Construction and the State Administration of Cultural Heritage. The book became a key reference material for the successful UNESCO nomination of Wutaishan in June 2009. In 2006, the School of Architecture continued fieldwork on Mount Wutai, this time under the leadership of He Congrong. The team members, among them graduate students Bao Zhiyu and Liu Yishi and ten bachelor students (Class of 2003), surveyed and measured the buildings at Pusading (Fig.14).

All of the measured drawings provide detailed information for future conservation and repair of the historical architecture on Mount Wutai, and contribute to a more comprehensive and systematic study of traditional Chinese architecture.

We owe special gratitude to the local administration, and I would like to thank the head of Mount Wutai Landscape Management and the director of Historical Records of Mount Wutai Religious Relics Bureau. Furthermore, I would like to express my thanks for the mental and spiritual support by Abbot LIN Hu of Pusading and Master Bao Gui. TANG Henglu from WANG Guixiang Studio was responsible for the revision and arrangement of the architectural drawings, together with his colleagues MAI Lin lin, SHAN Menglin, and HU Jingfu, and I would also like to thank them for their hard work.

Written by LIAO Huinong
Architecture History and Historic Preservation Research Institute, School of Architecture, Tsinghua University

Translated by Alexandra Harrer

2006 年清华大学建筑学院又派出教师和学生对五台山的其他寺庙建筑进行了测绘。2006 年 7 月由贺从容老师带队，研究生包志禹、刘亦师参加辅导，2003 级本科生 10 人前往五台山测绘菩萨顶寺庙（图 14）。这些测绘图都为今后的五台山寺庙建筑的保护和修缮提供了翔实的资料，并为古建筑研究打下坚实的基础。这几次测绘工作都得到了五台山风景管理区郑区长、五台山宗教文物局史册局长等当地许多同志的积极协助。同时五台山菩萨顶的林虎方丈、宝贵法师也予以大力支持和帮助，在此一并表示感谢。本册测绘图的修改和整理是由王贵祥老师工作室的唐恒鲁负责，由买琳琳、单梦林及胡竞芙具体完成。在此表示感谢。

清华大学建筑学院　建筑历史与文物保护研究所

廖慧农

图
版

Figure

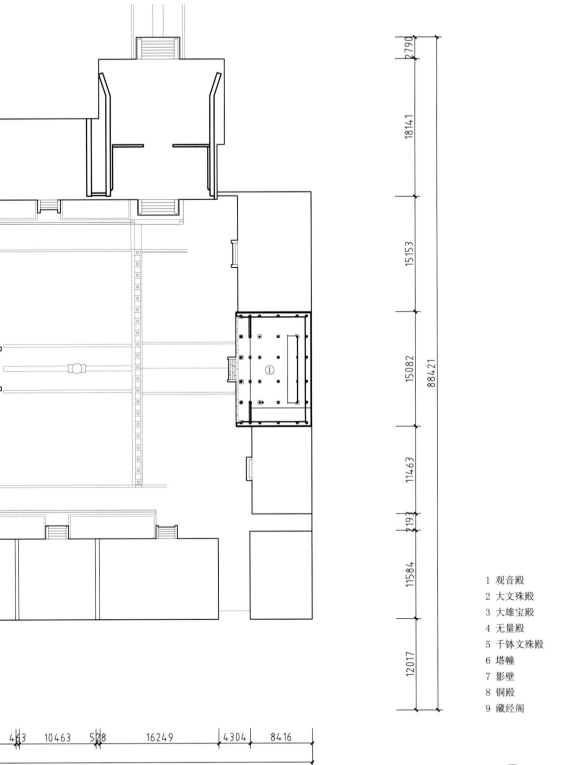

中国古建筑测绘大系·宗教建筑——五台山佛教建筑

020

1 观音殿
2 大文殊殿
3 大雄宝殿
4 无量殿
5 千钵文殊殿
6 塔幢
7 影壁
8 铜殿
9 藏经阁

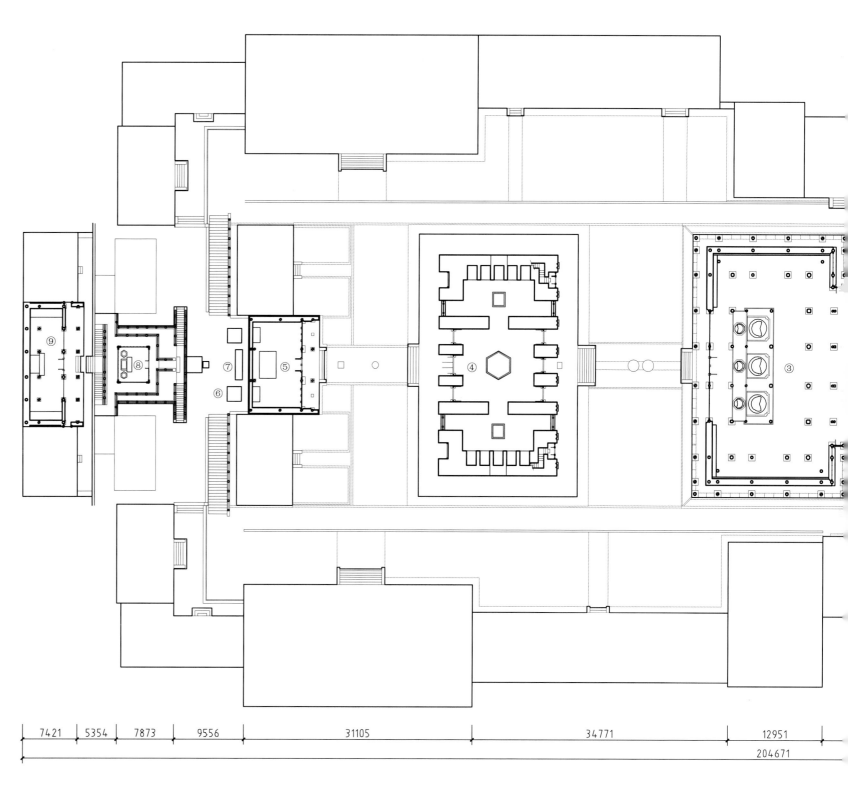

7421　5354　7873　9556　31105　34771　12951

204671

显通寺总平面图

Site plan of Xiantongsi

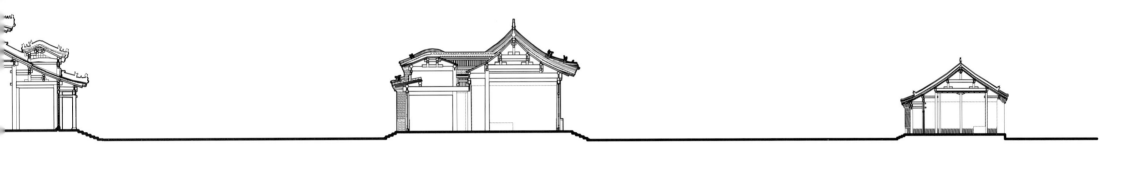

| 25561 | | 25508 | 17992 | 33880 | 10270 | |

337

0   5   15m

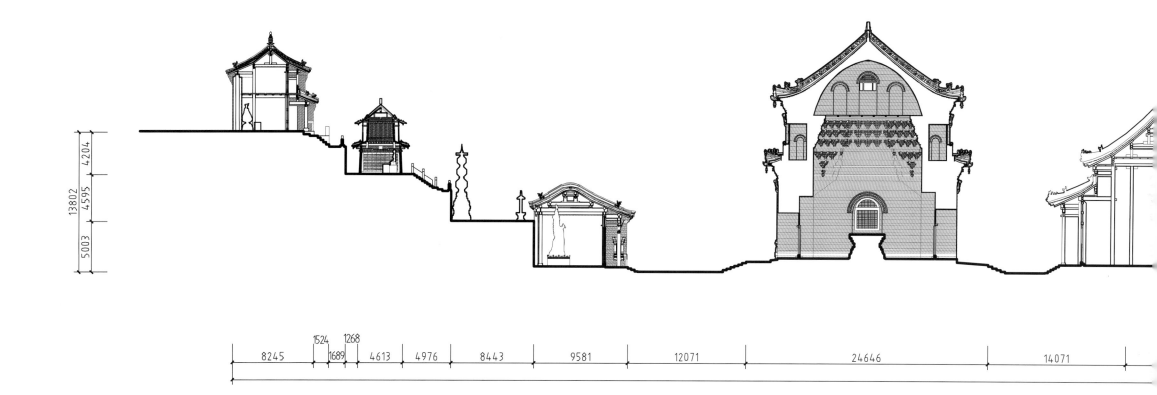

显通寺总剖面图

Site section of Xiantongsi

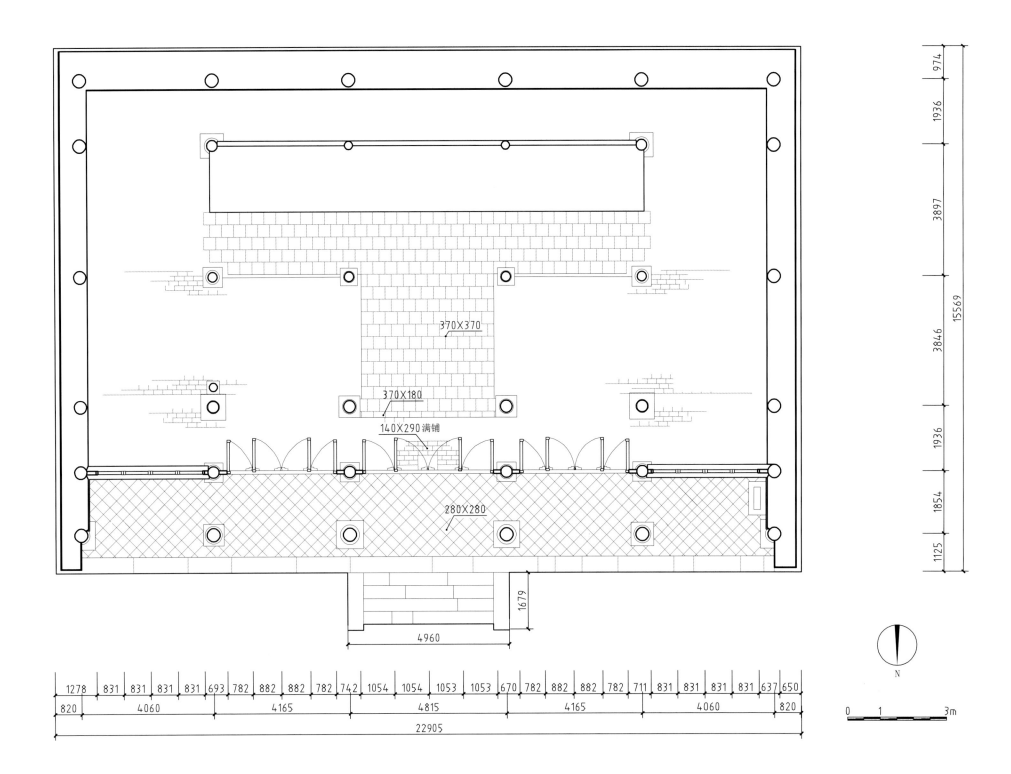

370×370

370×180

140×290满铺

280×280

974
1936
3897
3846
15569
1936
1854
1125

1679

4960

1278 831 831 831 831 693 782 882 882 782 742 1054 1054 1053 1053 670 782 882 882 782 711 831 831 831 831 637 650
820 4060 4165 4815 4165 4060 820
22905

N

0 1 3m

显通寺观音殿平面图
Plan of Guanyindian of Xiantongsi

正脊 10.853

戗脊旁四垄为瓦垄做法

戗脊上皮 5.351
滴水上皮 4.715
飞椽上皮 4.564

1.110

±0.000

820　4060　4165　4815　4170　4055　820

22868

0　1　3m

显通寺观音殿北立面图

North elevation of Guanyindian of Xiantongsi

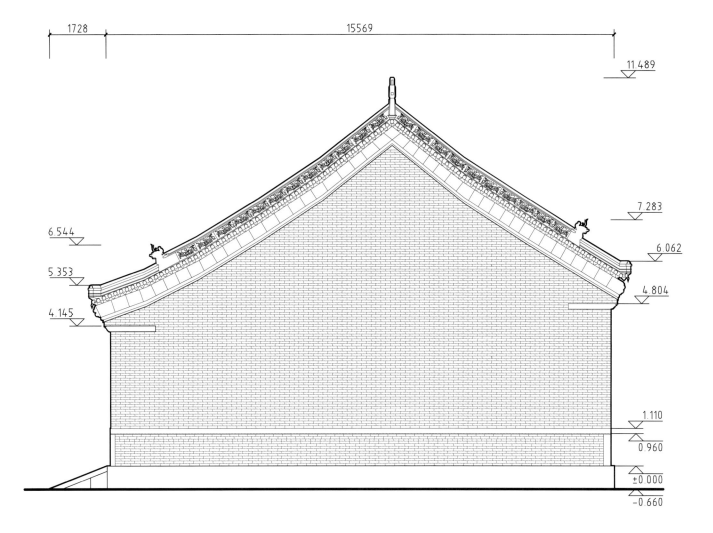

1728

15569

11.489

7.283

6.544

6.062

5.353

4.804

4.145

1.110

0.960

±0.000

-0.660

0　　1　　　　3m

显通寺观音殿西立面图

West elevation of Guanyindian of Xiantongsi

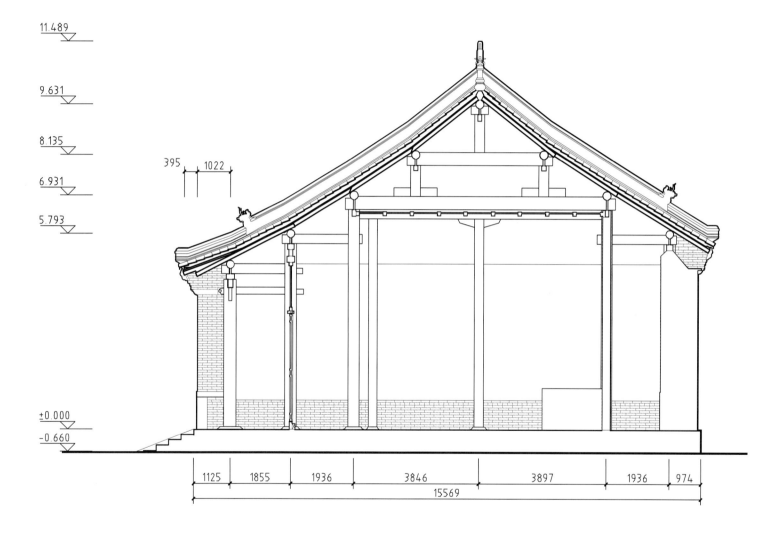

11.489

9.631

8.135

395   1022

6.931

5.793

±0.000
-0.660

1125   1855   1936   3846   3897   1936   974

15569

0   1   3m

显通寺观音殿明间横剖面图

Cross-section of central-bay of Guanyindian of Xiantongsi

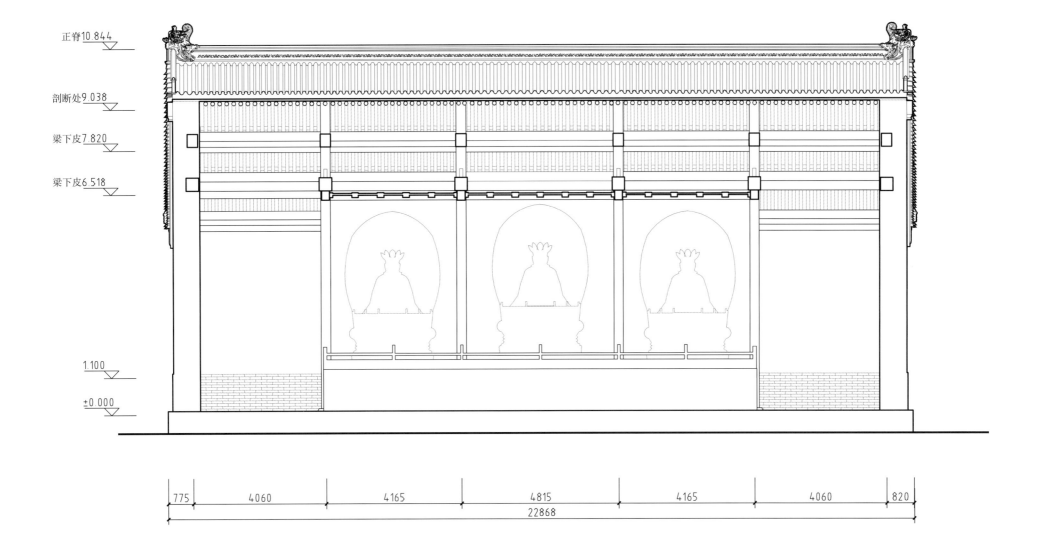

正脊10.844

剖断处9.038

梁下皮7.820

梁下皮6.518

1.100

±0.000

| 775 | 4060 | 4165 | 4815 | 4165 | 4060 | 820 |

22868

显通寺观音殿纵剖面图

Longitudinal section of Guanyindian of Xiantongsi

0　　1　　　　3m

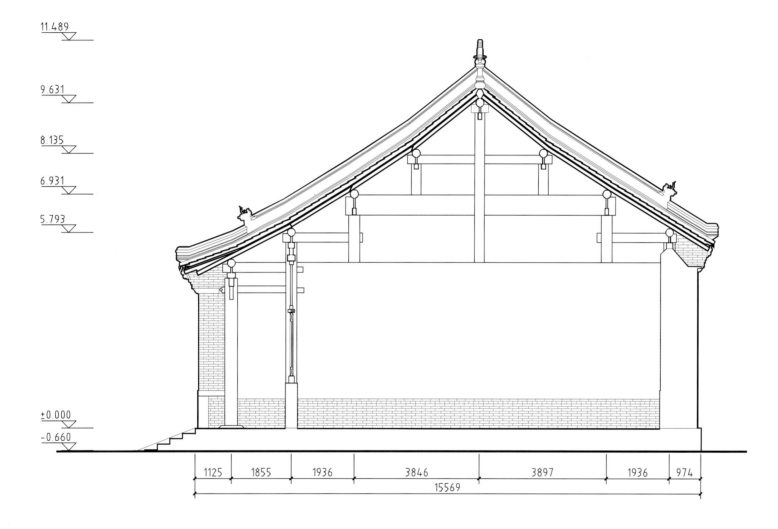

11.489

9.631

8.135

6.931

5.793

±0.000

-0.660

| 1125 | 1855 | 1936 | 3846 | 3897 | 1936 | 974 |

15569

0 1 3m

显通寺观音殿梢间横剖面图

Cross-section of second-to-last-bay of Guanyindian of Xiantongsi

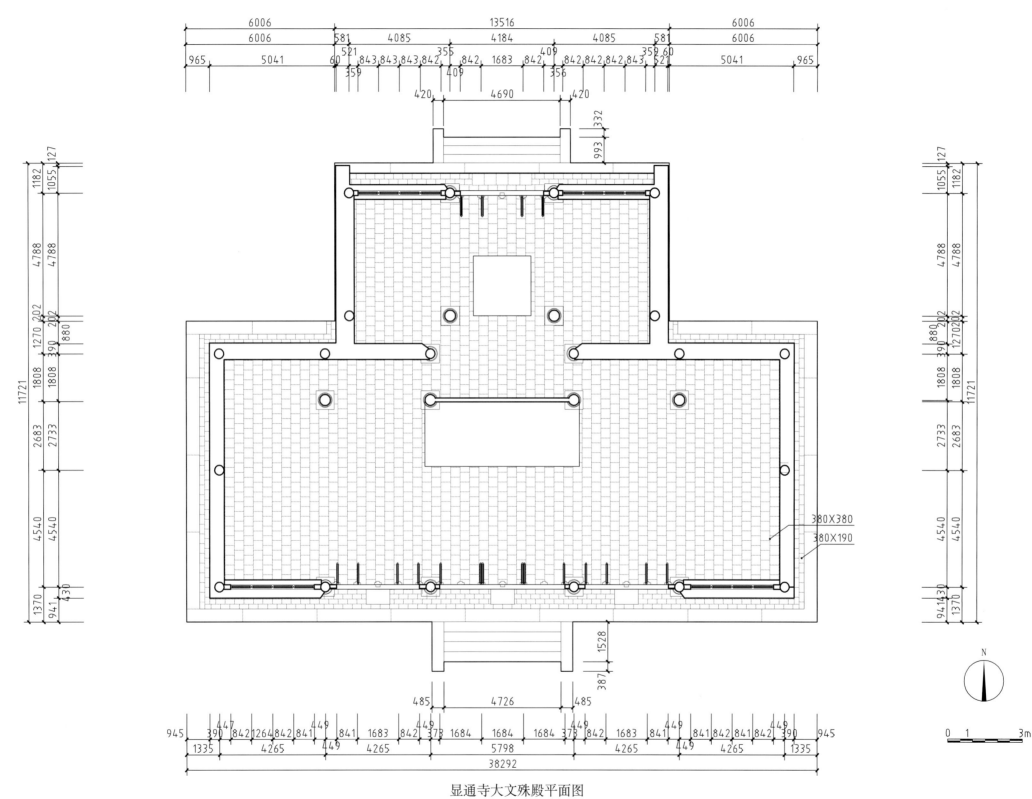

380×380
380×190

显通寺大文殊殿平面图
Plan of Da Wenshudian of Xiantongsi

N

0   1   3m

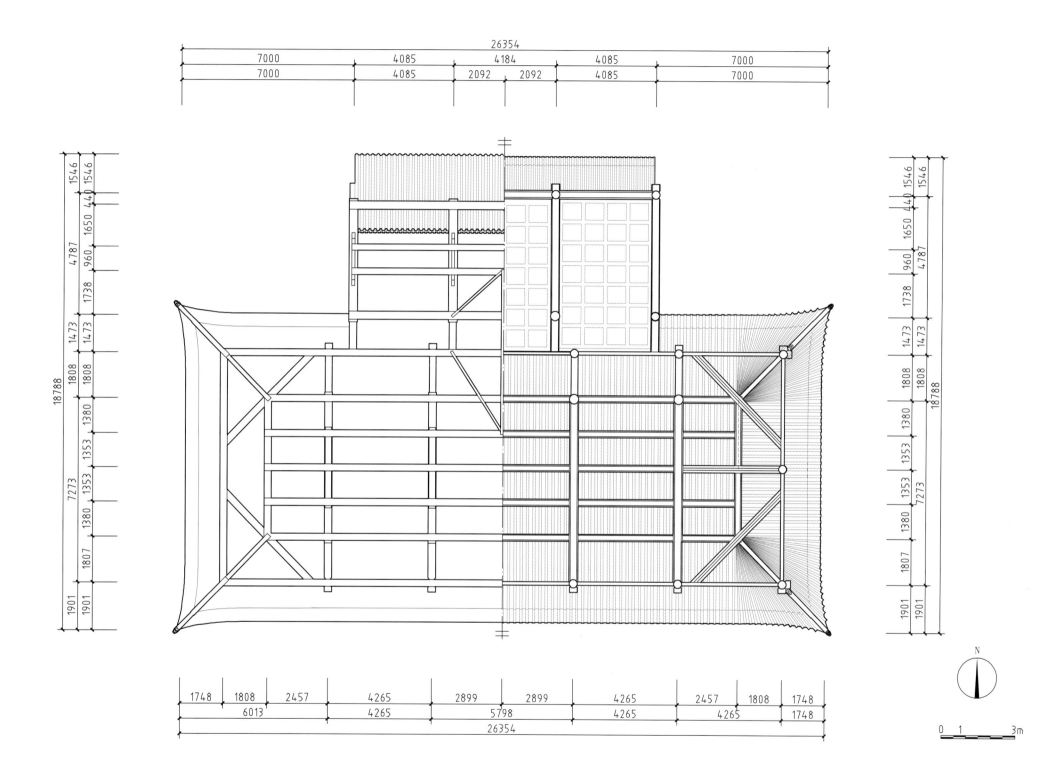

显通寺大文殊殿梁架俯仰视图

Plan of framework of Da Wenshudian of Xiantongsi as seen from above and below

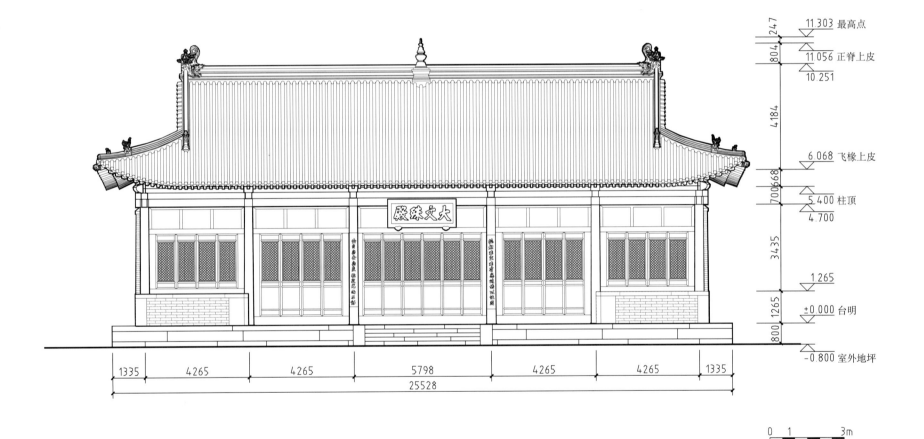

11.303 最高点

11.056 正脊上皮
10.251

6.068 飞椽上皮

5.400 柱顶
4.700

1.265

±0.000 台明

-0.800 室外地坪

247
804
4.184
668 700
3.435
1.265
800

1335　4265　4265　5798　4265　4265　1335

25528

显通寺大文殊殿南立面图
South elevation of Da Wenshudian of Xiantongsi

0　1　　3m

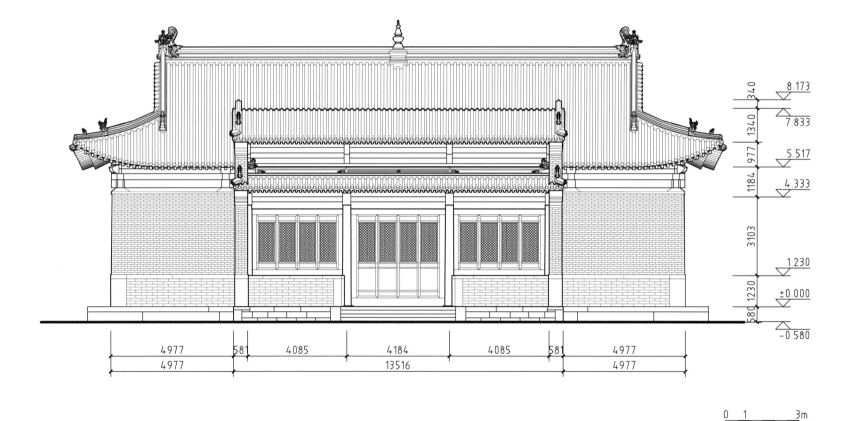

8.173
7.833
5.517
4.333
1.230
±0.000
-0.580

340
1340
977
1184
3103
1230
580

4977 | 581 | 4085 | 4184 | 4085 | 581 | 4977
4977 | 13516 | 4977

0 1 3m

显通寺大文殊殿北立面图

North elevation of Da Wenshudian of Xiantongsi

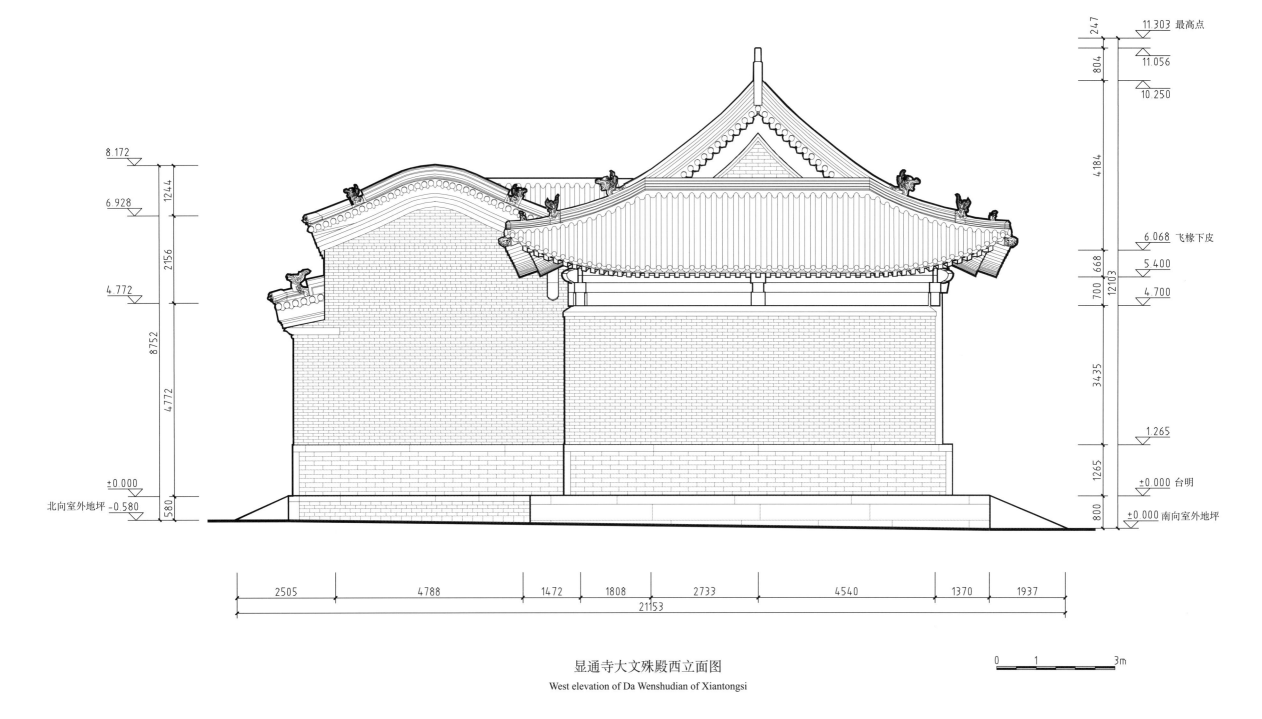

8.172

1244

6.928

2156

4.772

8752

4772

±0.000

580

北向室外地坪 -0.580

247 11.303 最高点

804 11.056

10.250

4184

6.068 飞椽下皮

668 5.400

700 4.700

1203

3435

1.265

1265 ±0.000 台明

800 ±0.000 南向室外地坪

2505 4788 1472 1808 2733 4540 1370 1937

21153

显通寺大文殊殿西立面图

West elevation of Da Wenshudian of Xiantongsi

0 1 3m

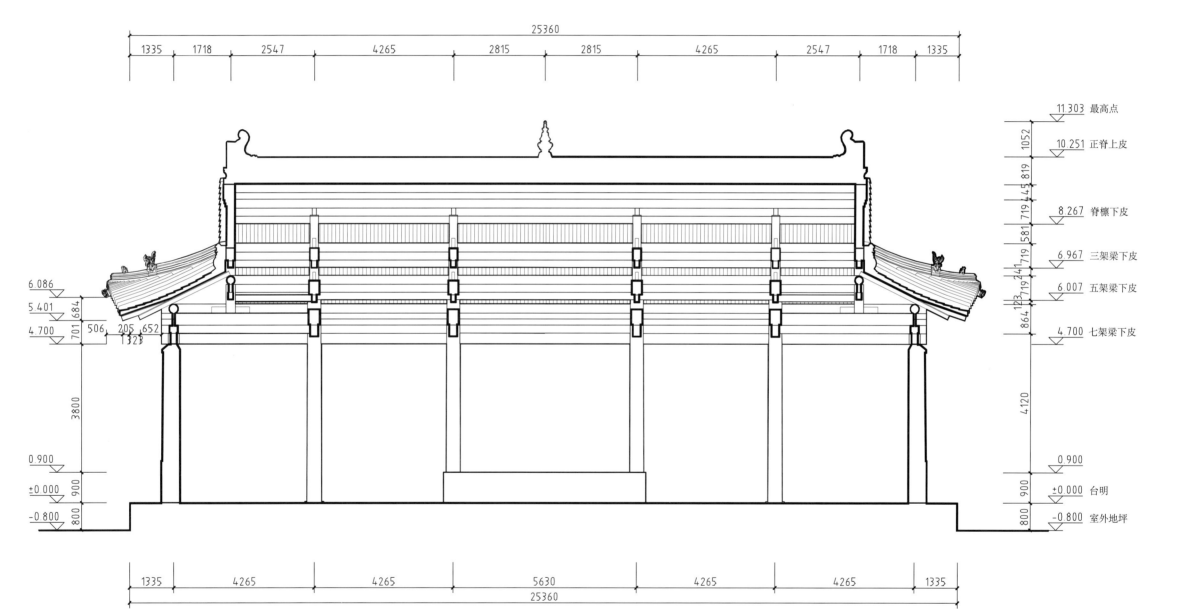

25360

1335 | 1718 | 2547 | 4265 | 2815 | 2815 | 4265 | 2547 | 1718 | 1335

11.303 最高点

10.251 正脊上皮

8.267 脊檩下皮

6.967 三架梁下皮

6.007 五架梁下皮

4.700 七架梁下皮

1052

819

719

581 719 241

719 123

864

4120

0.900

900

800

0.900

±0.000 台明

-0.800 室外地坪

6.086

5.401

4.700

701 684

506 205 652

1323

3800

0.900

±0.000

-0.800

900

800

1335 | 4265 | 4265 | 5630 | 4265 | 4265 | 1335

25360

0 1 3m

显通寺大文殊殿纵剖面图

Longitudinal section of Da Wenshudian of Xiantongsi

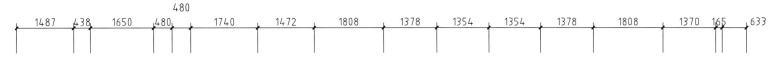

036

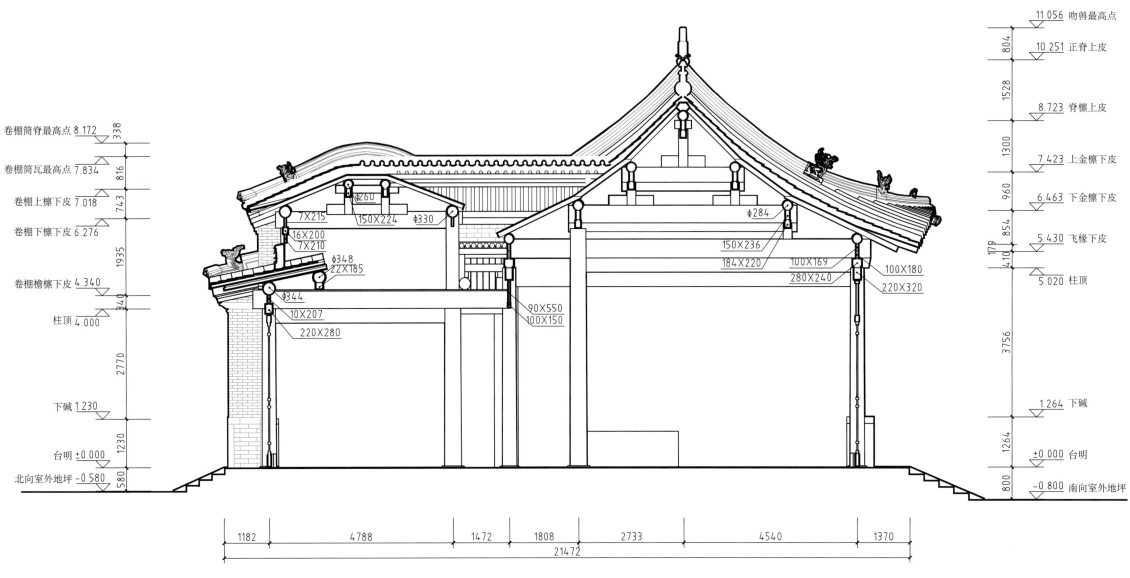

480

1487 438 1650 480 1740 1472 1808 1378 1354 1354 1378 1808 1370 165 633

11 056 吻兽最高点

804

10 251 正脊上皮

1528

8 723 脊檩上皮

1300

7 423 上金檩下皮

960

6 463 下金檩下皮

854

5 430 飞椽下皮

179

410

5 020 柱顶

3756

1 264 下碱

1264

±0.000 台明

800

-0.800 南向室外地坪

φ260
150×224
φ330
7×215
16×200
7×210
φ348
22×185
φ344
10×207
220×280
90×550
100×150

卷棚筒脊最高点 8 172   338
卷棚筒瓦最高点 7 834   816
卷棚上檩下皮 7 018   743
卷棚下檩下皮 6 276
                           1935
卷棚檐檩下皮 4 340   340
柱顶 4.000
                           2770
下碱 1.230
                           1230
台明±0.000
                           580
北向室外地坪 -0.580

φ284
150×236
184×220
100×169
280×240
100×180
220×320

1182 4788 1472 1808 2733 4540 1370
21472

0 1 3m

显通寺大文殊殿明间横剖面图

Cross-section of central-bay of Da Wenshudian of Xiantongsi

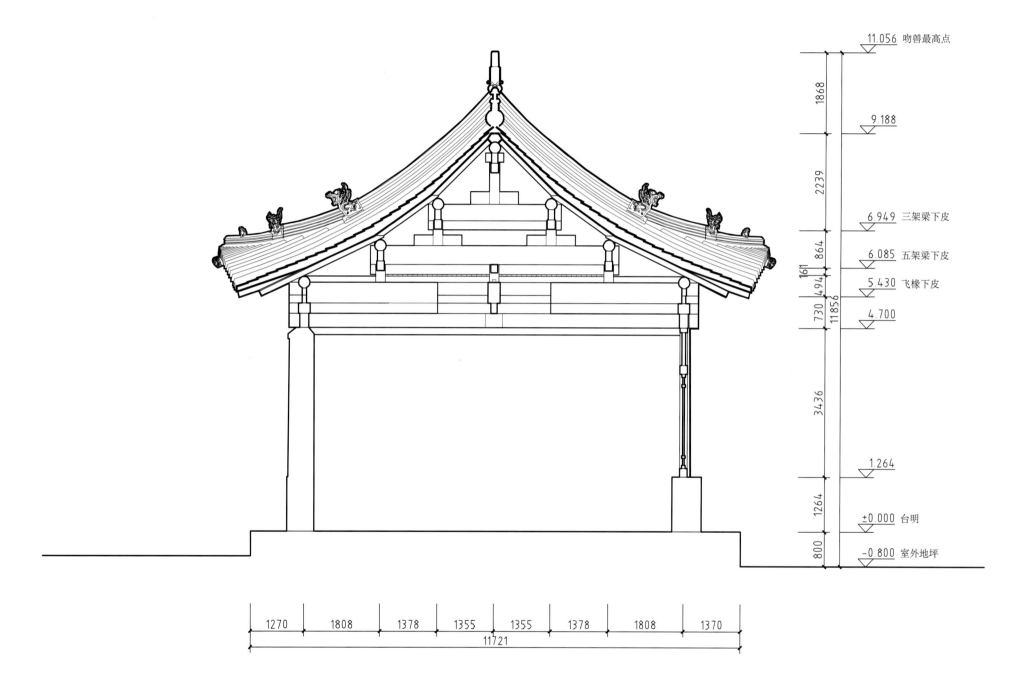

11.056 吻兽最高点

1868

9.188

2239

6.949 三架梁下皮

864

6.085 五架梁下皮

161

494

5.430 飞椽下皮

730

4.700

11856

3436

1.264

1264

±0.000 台明

800

-0.800 室外地坪

1270　1808　1378　1355　1355　1378　1808　1370

11721

显通寺大文殊殿梢间横剖面图

Cross-section of second-to-last-bay of Da Wenshudian of Xiantongsi

0　1　3m

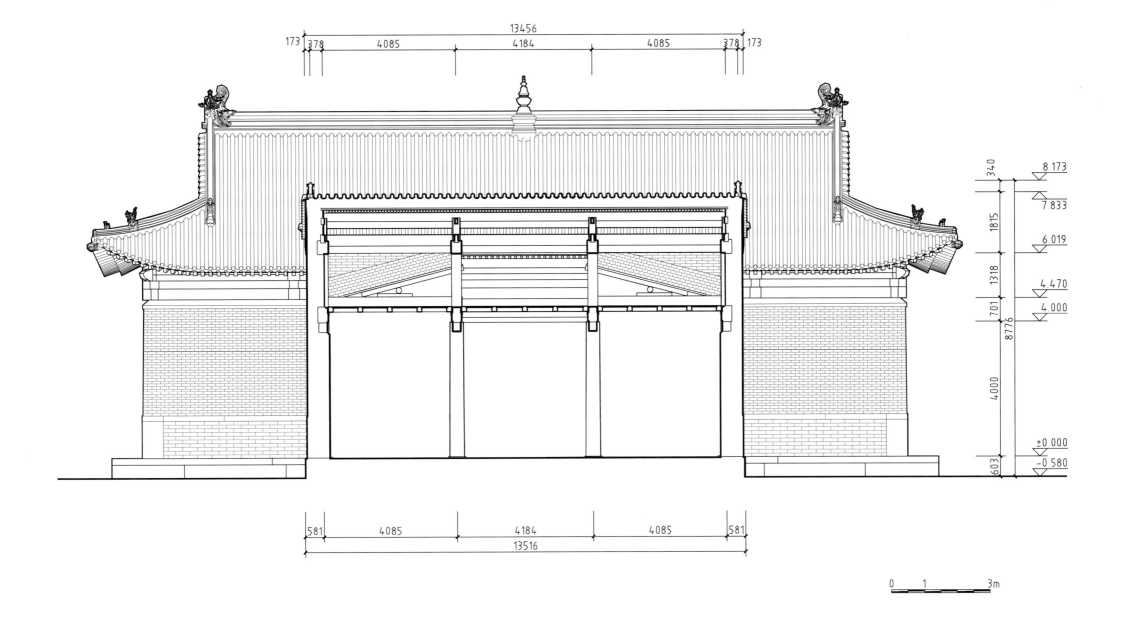

显通寺大文殊殿抱厦纵剖面图

Longitudinal section of *baoxia* of Da Wenshudian of Xiantongsi

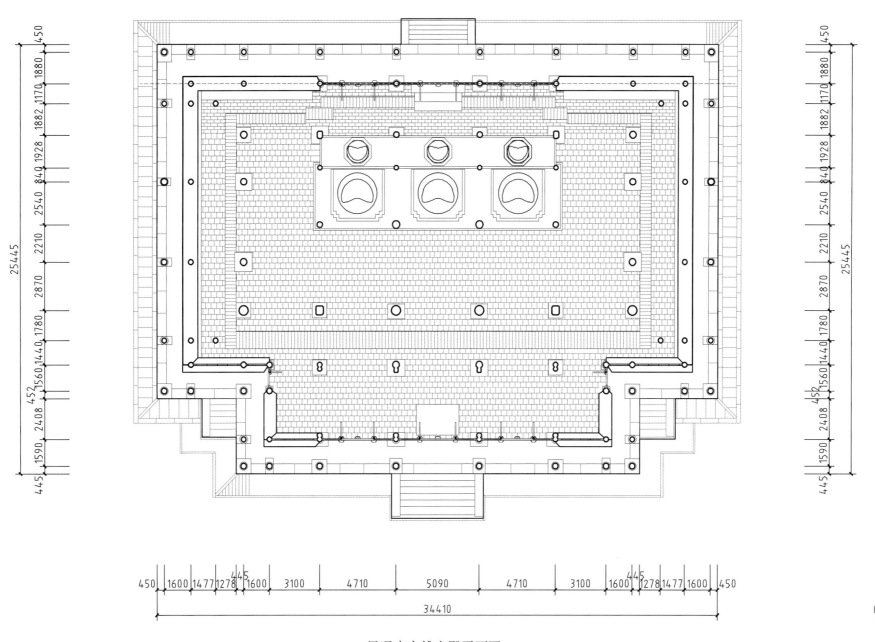

显通寺大雄宝殿平面图
Plan of Daxiong baodian of Xiantongsi

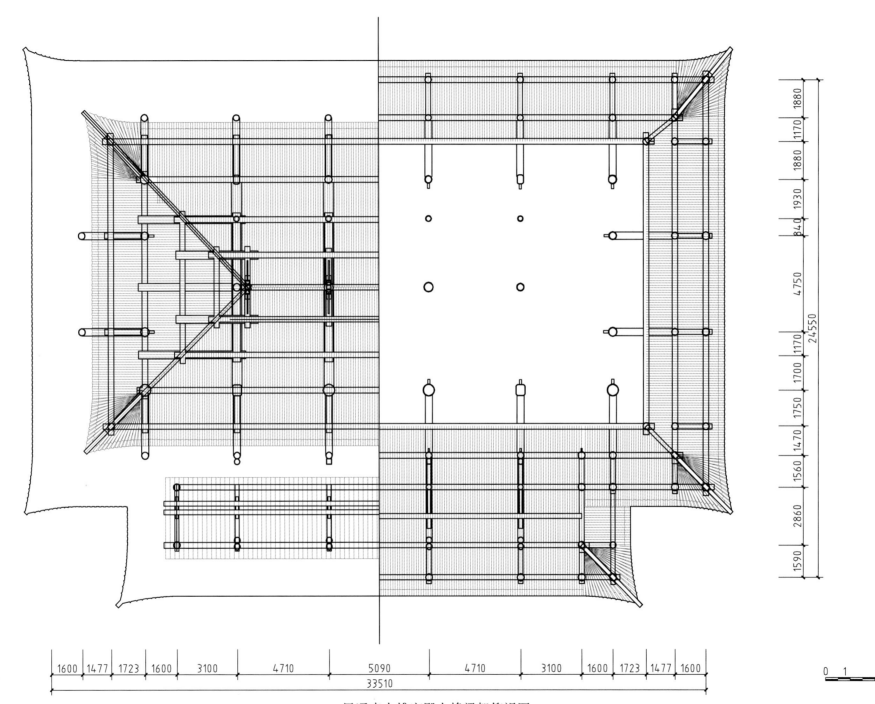

显通寺大雄宝殿上檐梁架仰视图

Plan of upper eaves framework of Daxiong baodian of Xiantongsi as seen from below

0 1 3m

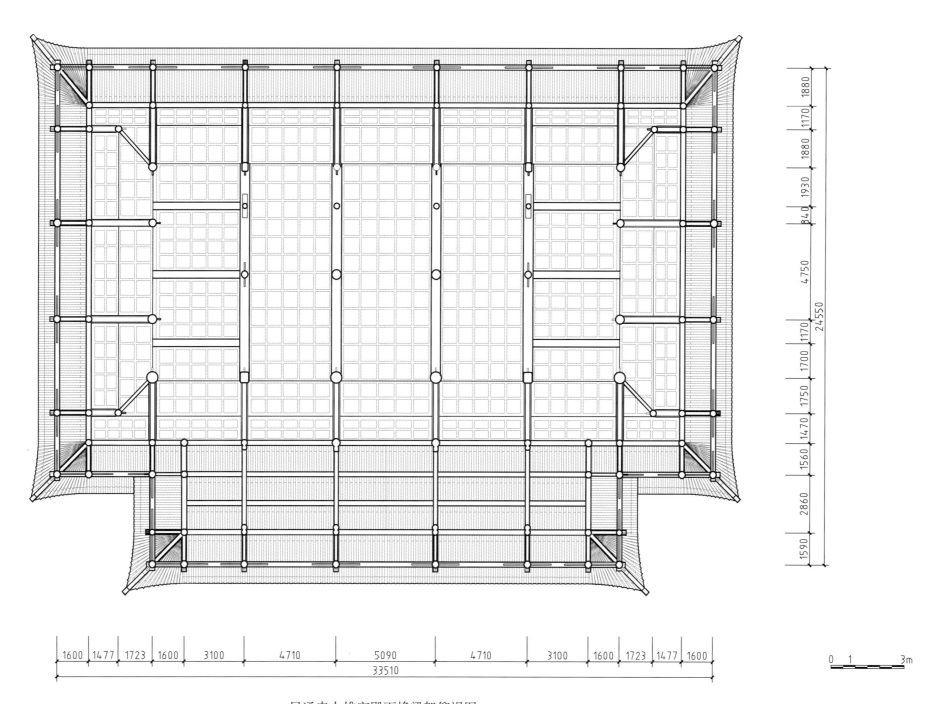

显通寺大雄宝殿下檐梁架仰视图

Plan of lower eaves framework of Daxiong baodian of Xiantongsi as seen from below

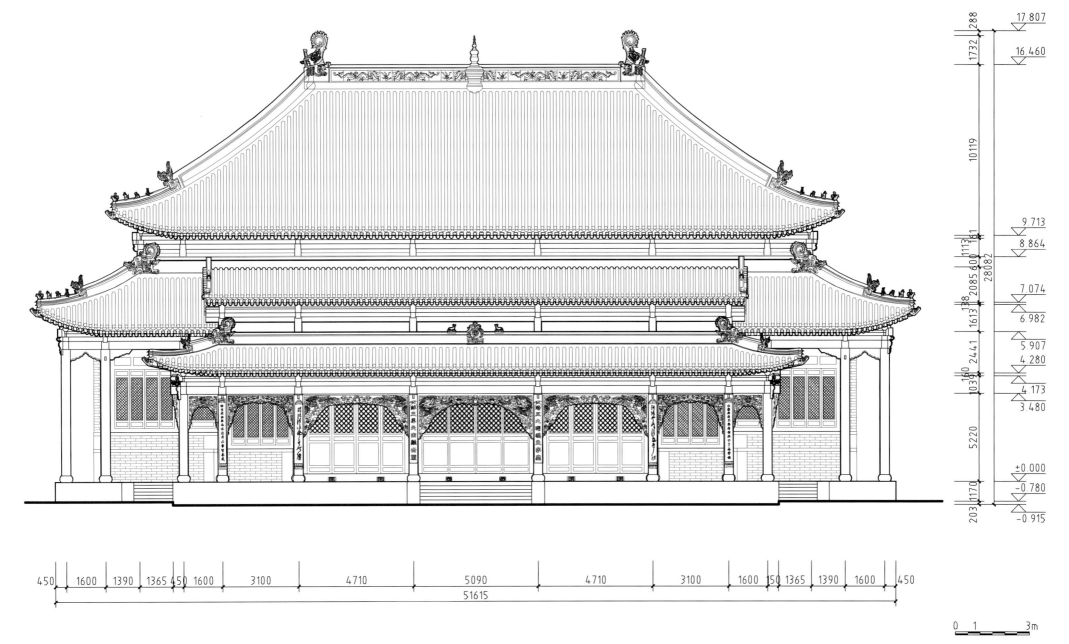

17.807
16.460
9.713
8.864
7.074
6.982
5.907
4.280
4.173
3.480
±0.000
-0.780
-0.915

288
1732
10119
161
1113
600
138
2085
1613
160
1039
5220
1170
203
28082
2441

450 | 1600 | 1390 | 1365 450 | 1600 | 3100 | 4710 | 5090 | 4710 | 3100 | 1600 150 1365 | 1390 | 1600 | 450
51615

0 1   3m

显通寺大雄宝殿南立面图

South elevation of Daxiong baodian of Xiantongsi

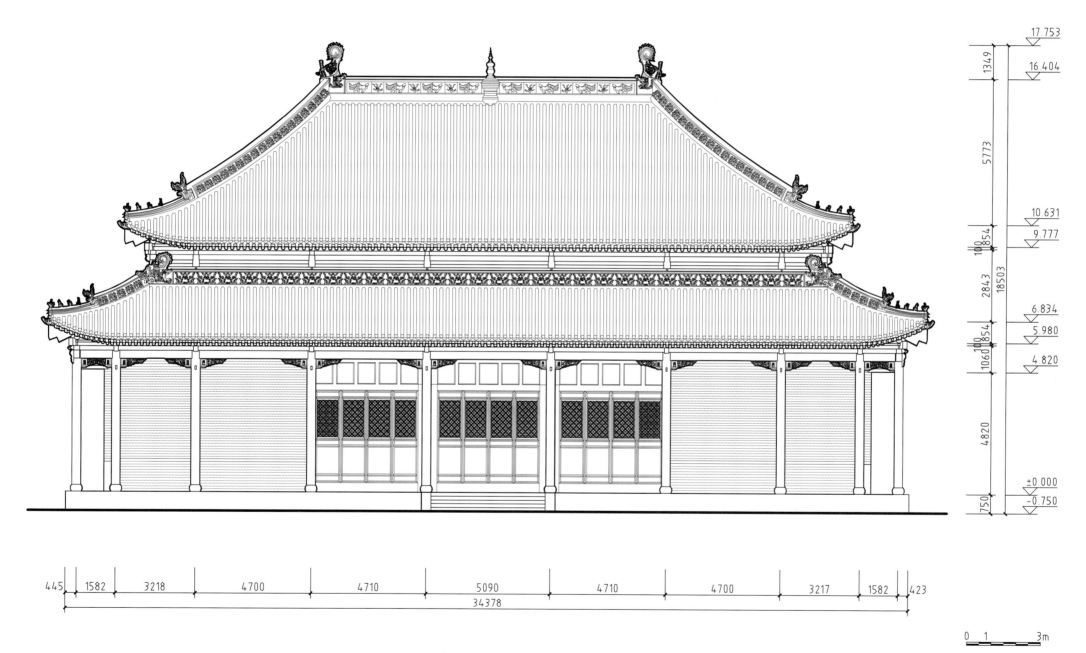

17.753

16.404

1349

5773

10.631

9.777

100
854

2843

18503

6.834

5.980

100
854

106

4.820

4820

±0.000

−0.750

750

445 | 1582 | 3218 | 4700 | 4710 | 5090 | 4710 | 4700 | 3217 | 1582 | 423

34378

0   1   3m

显通寺大雄宝殿北立面图

North elevation of Daxiong baodian of Xiantongsi

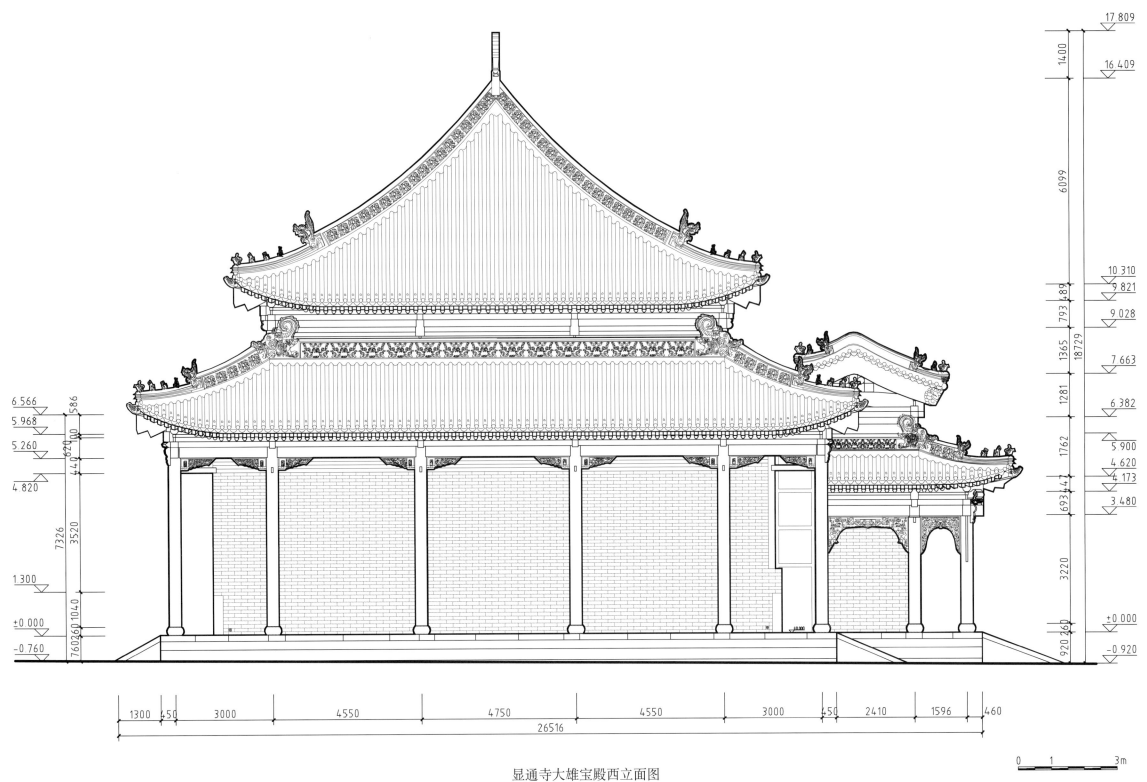

17.809

16.409

1400

6099

10.310
9.821

9.028

7.663

6.382

5.900

4.620
4.173

3.480

±0.000

-0.920

793 489

18729

1365

1281

1762

693 447

3220

920 260

6.566

5.968

5.260

4.820

1.300

±0.000

-0.760

586

620 100

440 100

7326

3520

1040

760 260

1300 450 3000 4550 4750 4550 3000 450 2410 1596 460

26516

0   1   3m

显通寺大雄宝殿西立面图

West elevation of Daxiong baodian of Xiantongsi

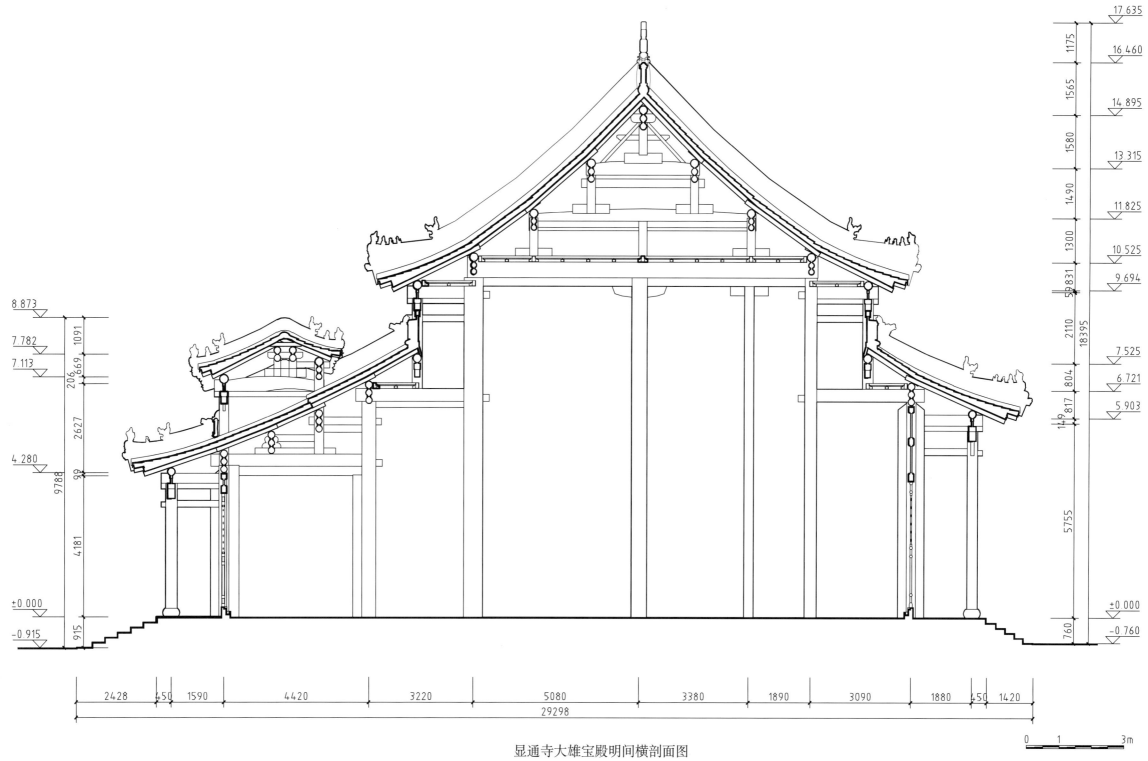

显通寺大雄宝殿明间横剖面图

Cross-section of central-bay of Daxiong baodian of Xiantongsi

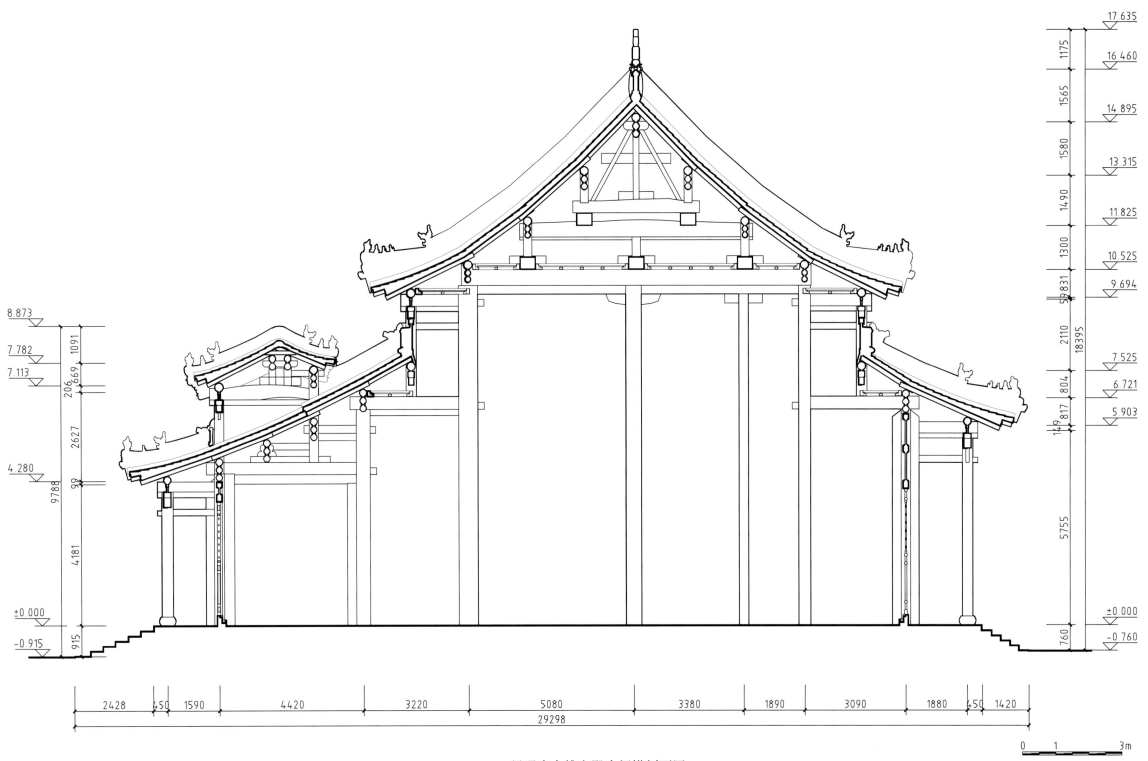

17.635
1175
16.460
1565
14.895
1580
13.315
1490
11.825
1300
10.525
831
9.694
2110
18395
7.525
804
6.721
817
5.903
5755
±0.000
760
-0.760

8.873
1091
7.782
206 669
7.113
2627
4.280
99
9788
4.280
4181
±0.000
915
-0.915

2428 450 1590 4420 3220 5080 3380 1890 3090 1880 450 1420
29298

显通寺大雄宝殿次间横剖面图
Cross-section of side-bay of Daxiong baodian of Xiantongsi

0 1 3m

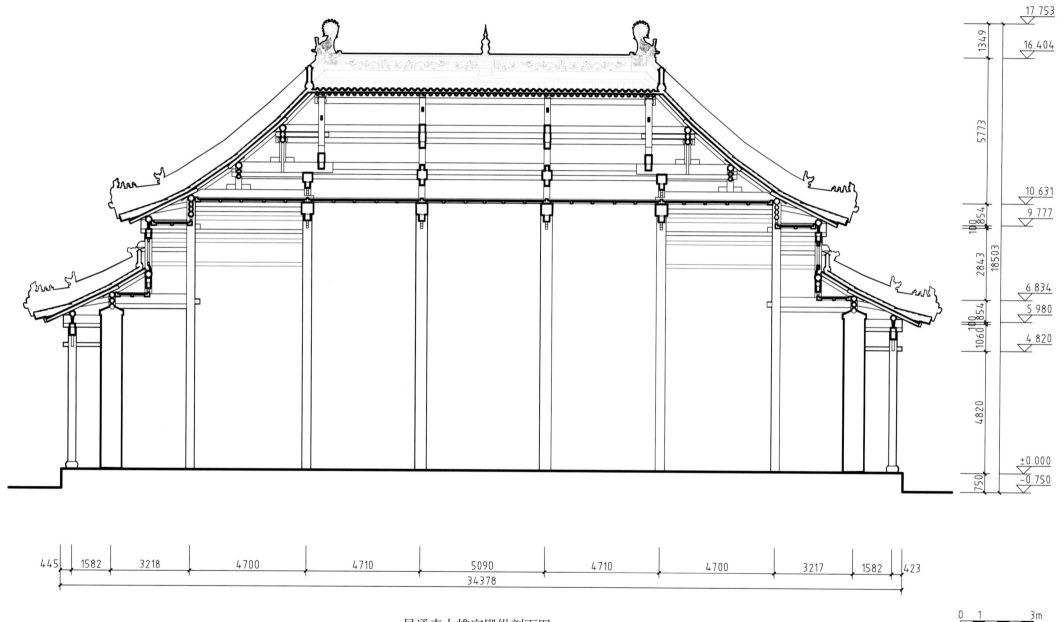

17.753

134.9

16.404

5773

10.631

100 854

9.777

18503

2843

6.834

100 854

5.980

1060

4.820

4820

±0.000

750

−0.750

445 | 1582 | 3218 | 4700 | 4710 | 5090 | 4710 | 4700 | 3217 | 1582 | 423

34378

0   1   3m

显通寺大雄宝殿纵剖面图

Longitudinal section of Daxiong baodian of Xiantongsi

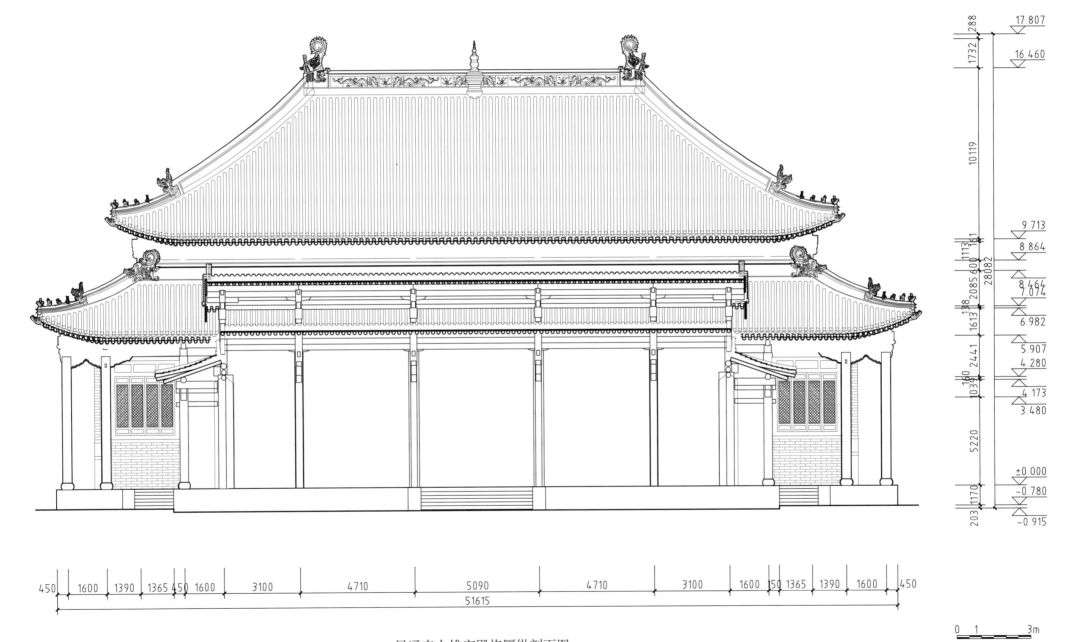

17.807
16.460
9.713
8.864
8.464
7.094
6.982
5.907
4.280
4.173
3.480
±0.000
-0.780
-0.915

288
1732
10119
151
1113
600
2085
38
1613
2441
160
1039
5220
28082
1170
203

450 | 1600 | 1390 | 1365 | 450 | 1600 | 3100 | 4710 | 5090 | 4710 | 3100 | 1600 | 150 | 1365 | 1390 | 1600 | 450
51615

0 1 3m

显通寺大雄宝殿抱厦纵剖面图

Longitudinal section of *baosha* of Daxiong baodian of Xiantongsi

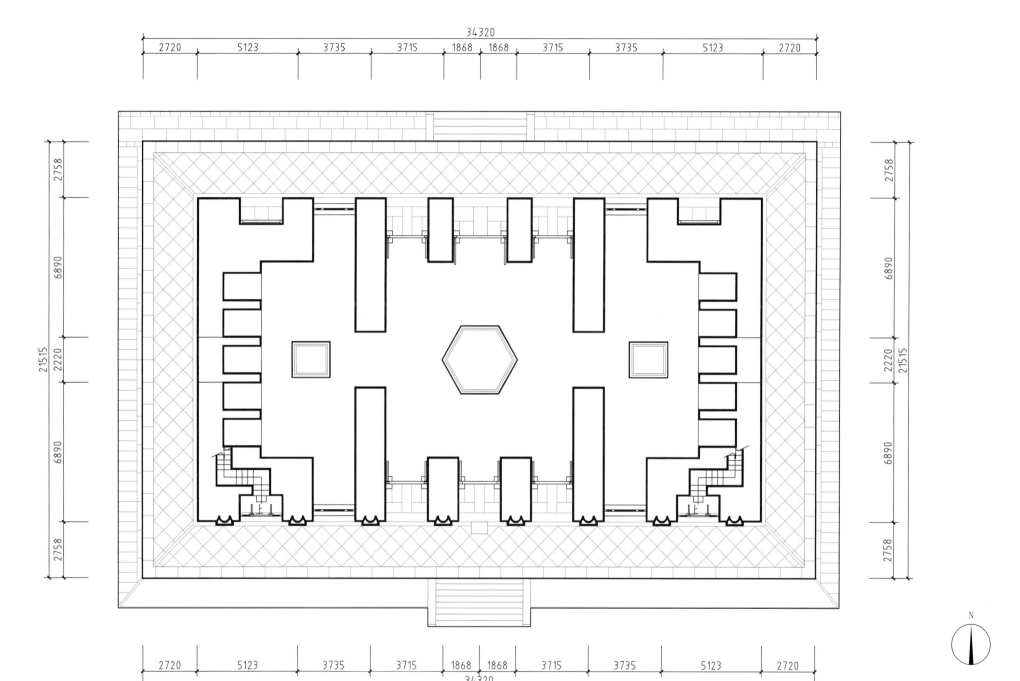

34320
2720 5123 3735 3715 1868 1868 3715 3735 5123 2720

2758
6890
2220 21515
6890
2758

2758
6890
2220 21515
6890
2758

2720 5123 3735 3715 1868 1868 3715 3735 5123 2720
34320

N

0 1 3m

显通寺无量殿一层平面图
Plan of first floor of Wuliangdian of Xiantongsi

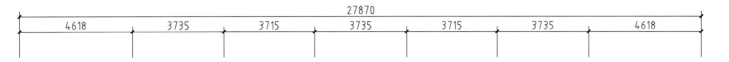

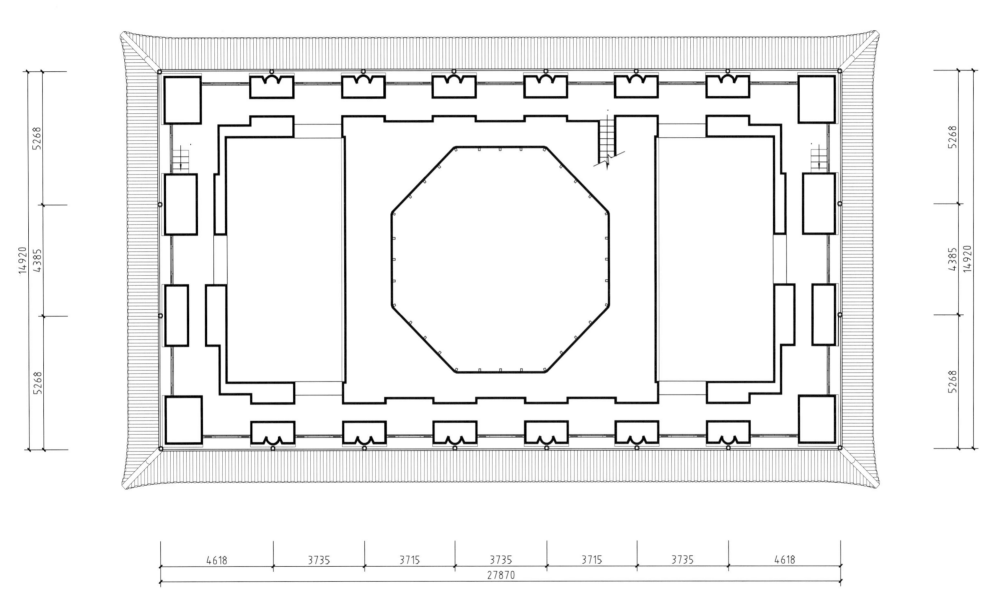

27870
4618　3735　3715　3735　3715　3735　4618

5268　4385　5268
14920

5268　4385　5268
14920

4618　3735　3715　3735　3715　3735　4618
27870

0　1　3m

显通寺无量殿二层平面图
Plan of second floor of Wuliangdian of Xiantongsi

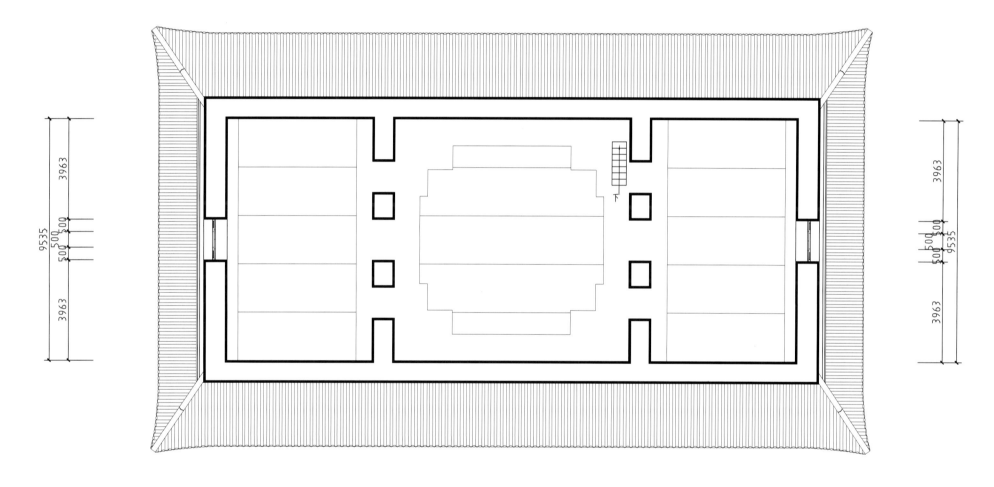

显通寺无量殿三层平面图

Plan of third floor of Wuliangdian of Xiantongsi

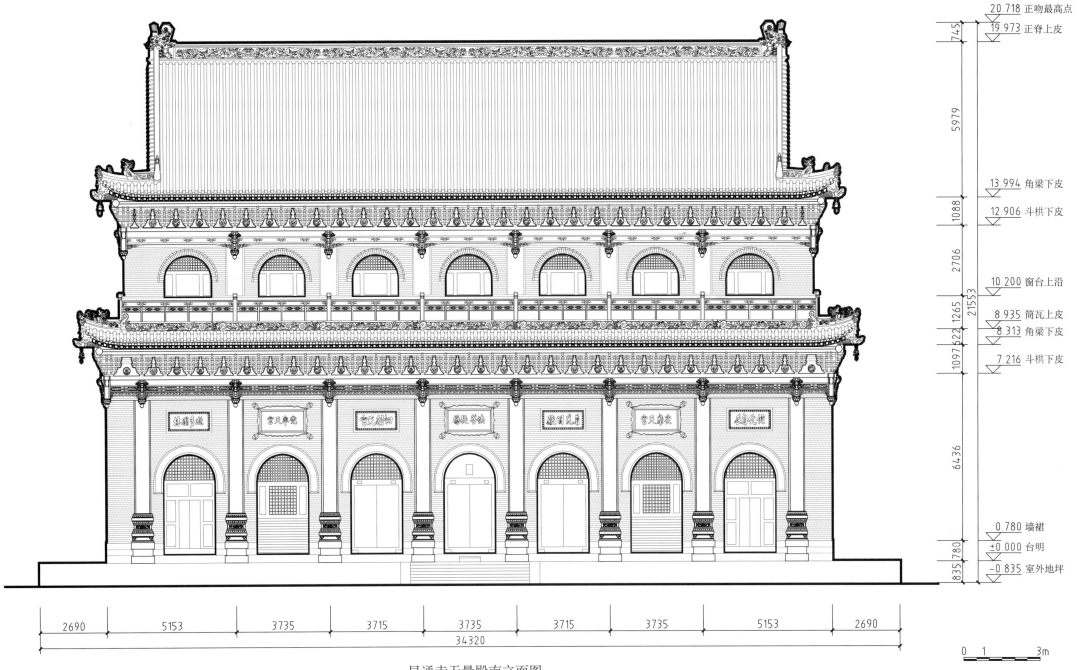

20.718 正吻最高点
19.973 正脊上皮
745
5979
13.994 角梁下皮
12.906 斗栱下皮
1088
2706
10.200 窗台上沿
21553
1265
8.935 筒瓦上皮
8.313 角梁下皮
522
1097
7.216 斗栱下皮
6436
0.780 墙裙
±0.000 台明
780
-0.835 室外地坪
835

2690 5153 3735 3715 3735 3715 3735 5153 2690
34320

0 1 3m

显通寺无量殿南立面图
South elevation of Wuliangdian of Xiantongsi

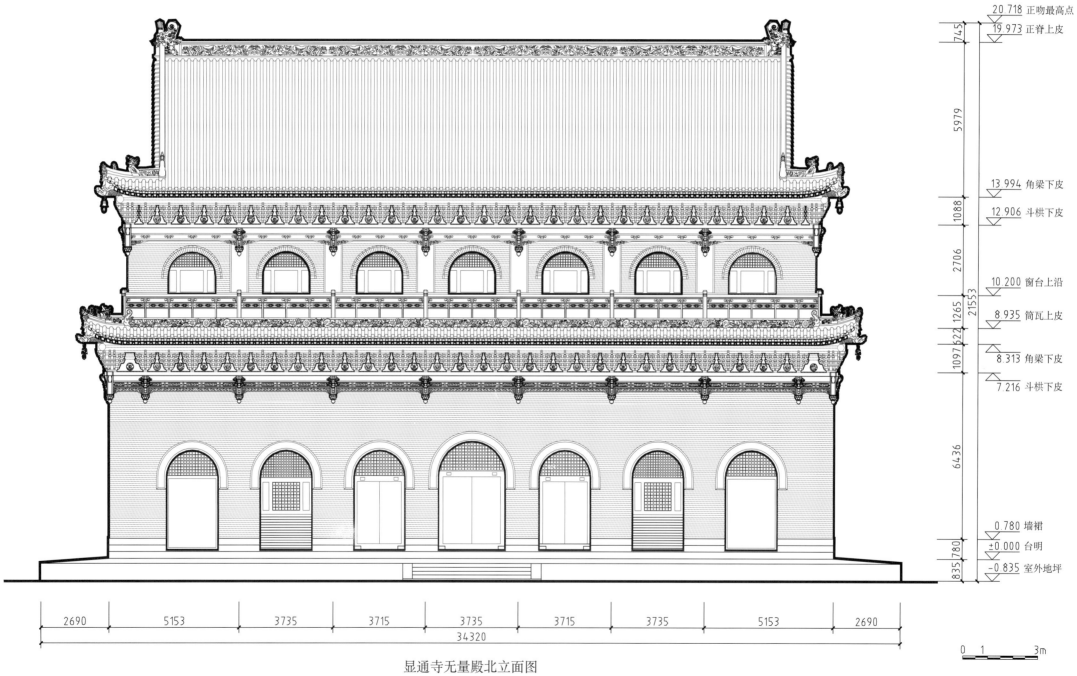

20 718 正吻最高点
19 973 正脊上皮
13 994 角梁下皮
12 906 斗栱下皮
10 200 窗台上沿
8 935 筒瓦上皮
8 313 角梁下皮
7 216 斗栱下皮
0 780 墙裙
±0 000 台明
-0 835 室外地坪

745
5979
1088
2706
1265
622
1097
6436
780
835
21553

2690　5153　3735　3715　3735　3715　3735　5153　2690
34320

0　1　3m

显通寺无量殿北立面图
North elevation of Wuliangdian of Xiantongsi

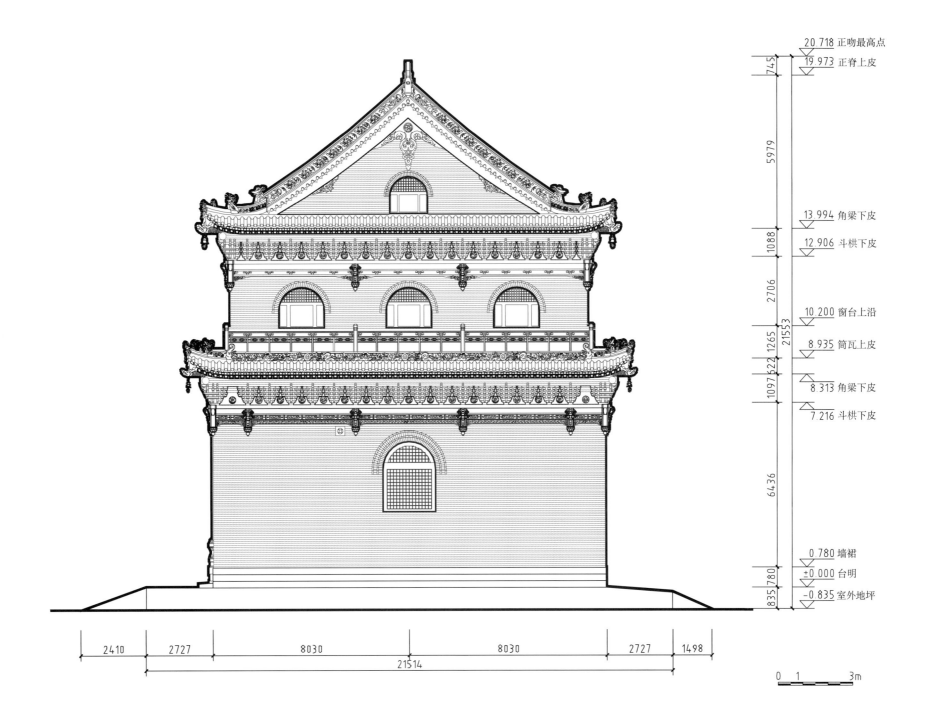

20.718 正吻最高点
745
19.973 正脊上皮

5979

13.994 角梁下皮
1088
12.906 斗栱下皮

2706

10.200 窗台上沿

1265
8.935 筒瓦上皮
21553
522
1097
8.313 角梁下皮

7.216 斗栱下皮

6436

0.780 墙裙
780
±0.000 台明
835
-0.835 室外地坪

2410   2727   8030   8030   2727   1498
21514

0 1   3m

显通寺无量殿东立面图
East elevation of Wuliangdian of Xiantongsi

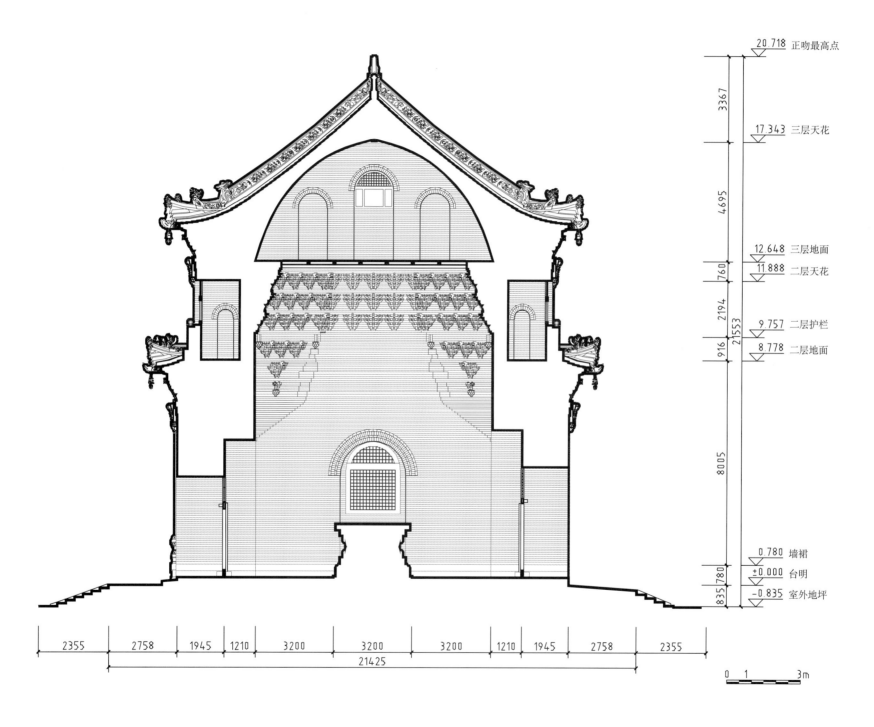

20.718 正吻最高点

3367

17.343 三层天花

4695

12.648 三层地面
11.888 二层天花

760

2194

9.757 二层护栏

916

8.778 二层地面

21553

8005

0.780 墙裙
±0.000 台明

780

-0.835 室外地坪

835

2355　2758　1945　1210　3200　3200　3200　1210　1945　2758　2355

21425

0　1　　　3m

显通寺无量殿横剖面图

Cross-section of Wuliangdian of Xiantongsi

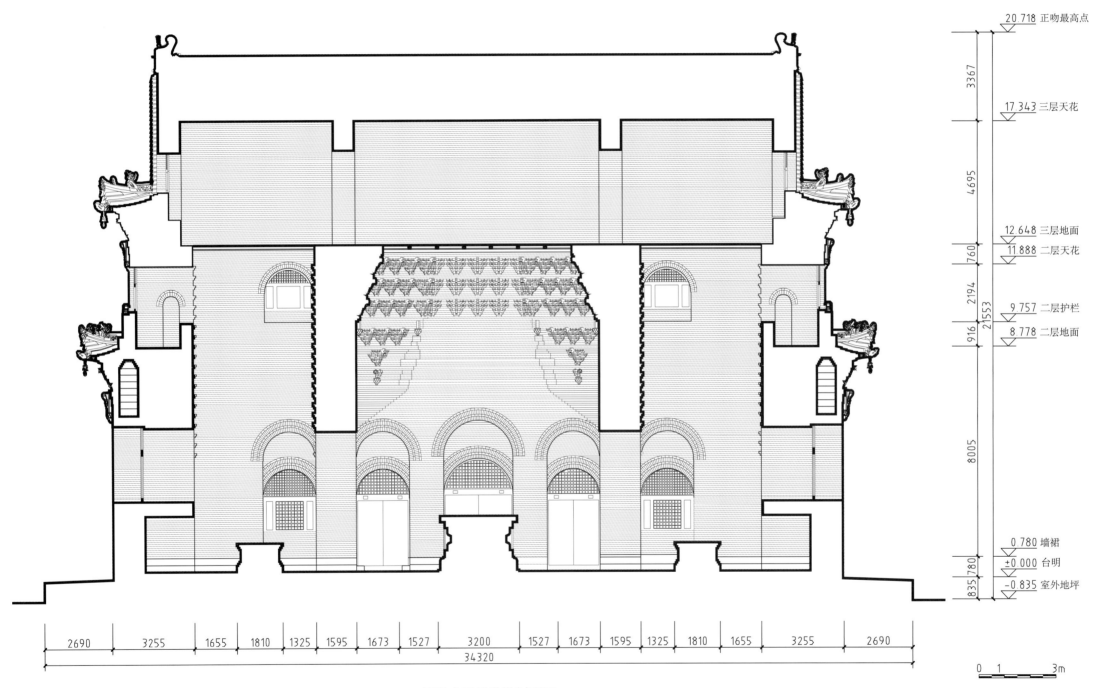

20.718 正吻最高点

17.343 三层天花

12.648 三层地面

11.888 二层天花

9.757 二层护栏

8.778 二层地面

0.780 墙裙

±0.000 台明

-0.835 室外地坪

3367

4695

760

2194

916

2155 3

8005

780

835

2690　3255　1655　1810　1325　1595　1673　1527　3200　1527　1673　1595　1325　1810　1655　3255　2690

34320

0　1　3m

显通寺无量殿纵剖面图
Longitudinal section of Wuliangdian of Xiantongsi

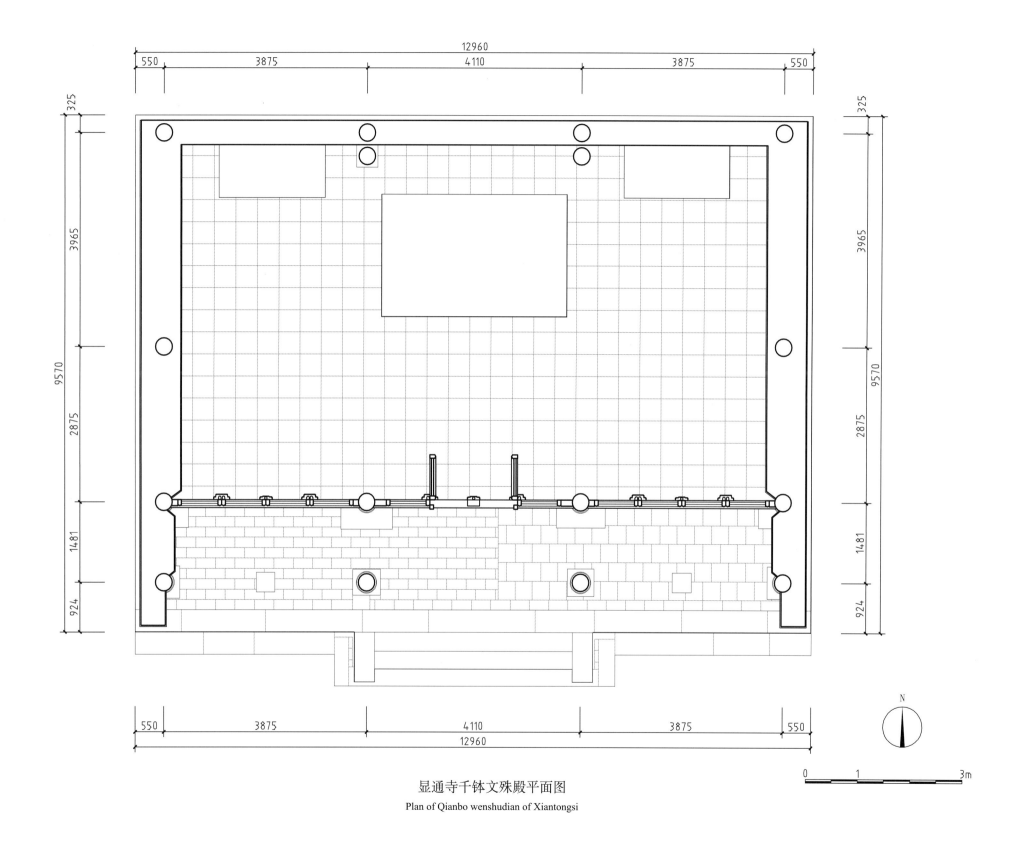

显通寺千钵文殊殿平面图
Plan of Qianbo wenshudian of Xiantongsi

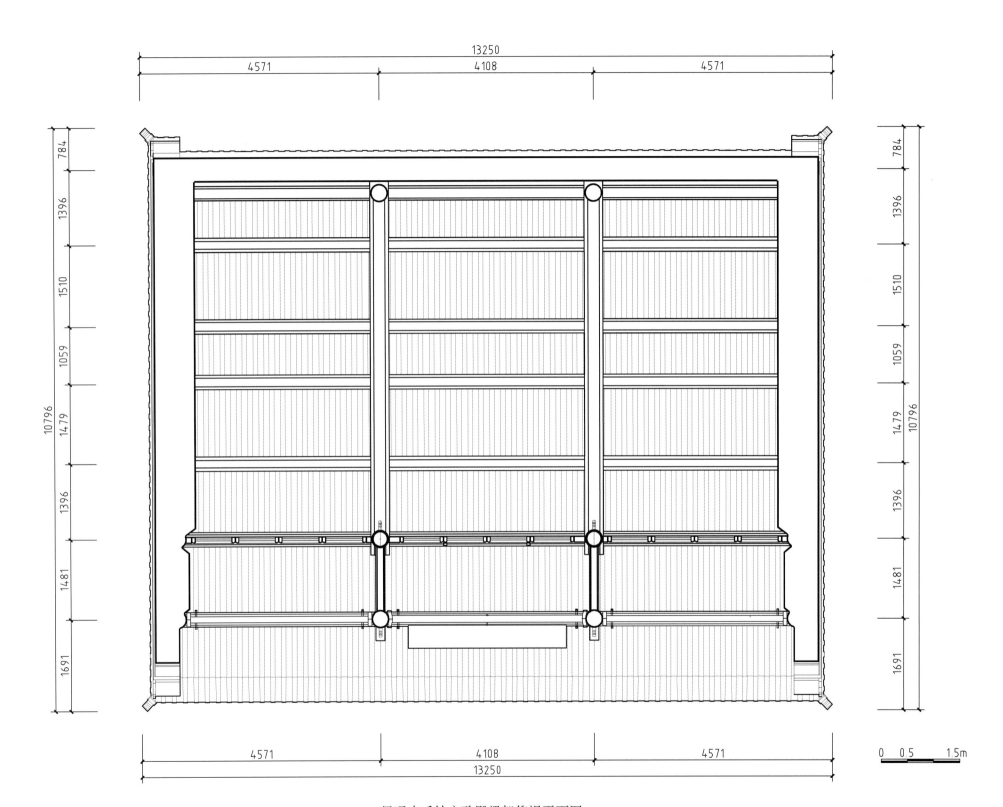

显通寺千钵文殊殿梁架仰视平面图

Plan of framework of Qianbo wenshudian of Xiantongsi as seen from below

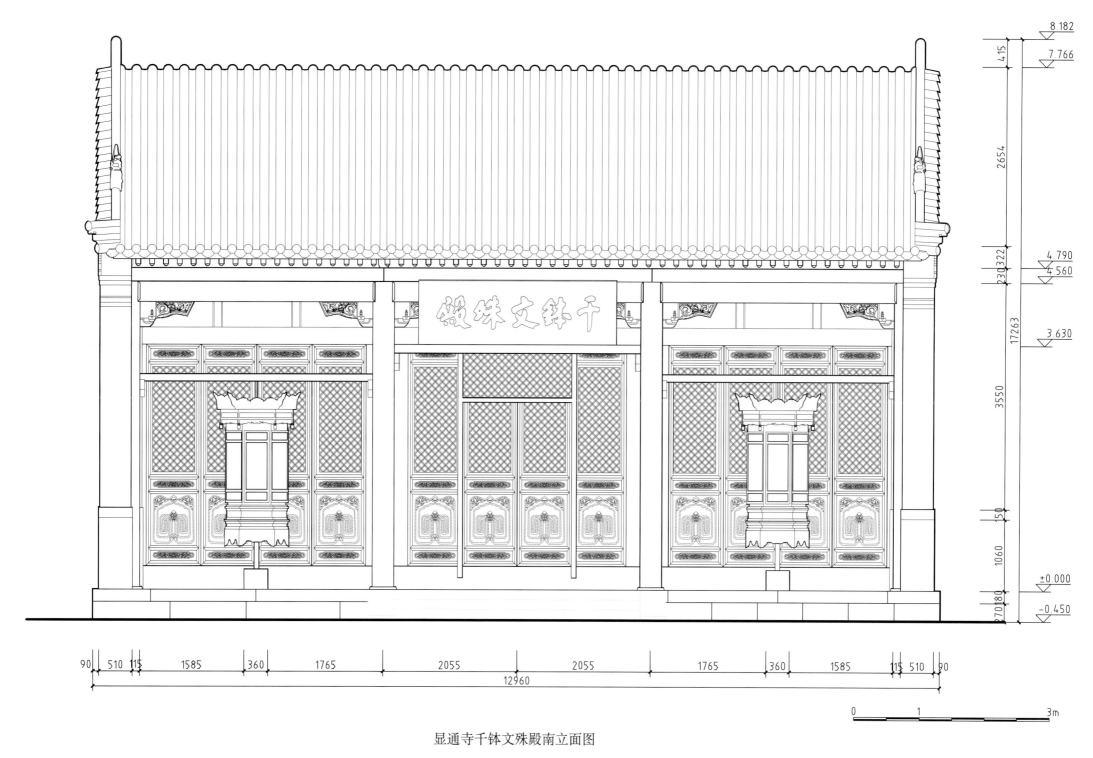

显通寺千钵文殊殿南立面图

South elevation of Qianbo wenshudian of Xiantongsi

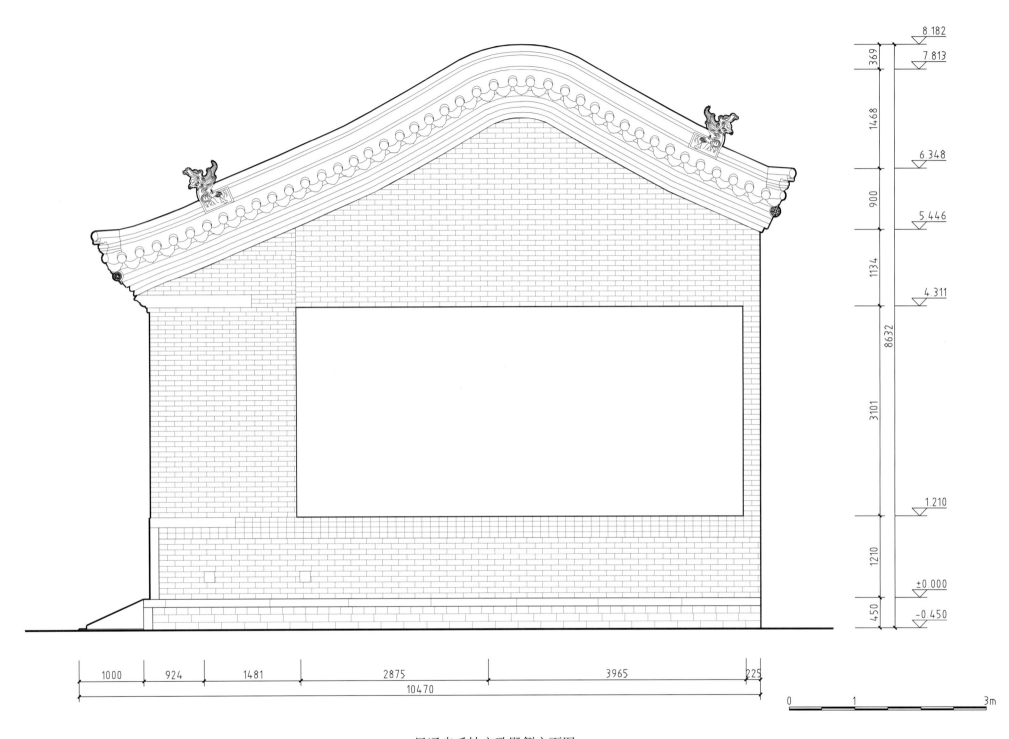

显通寺千钵文殊殿侧立面图
Side elevation of Qianbo wenshudian of Xiantongsi

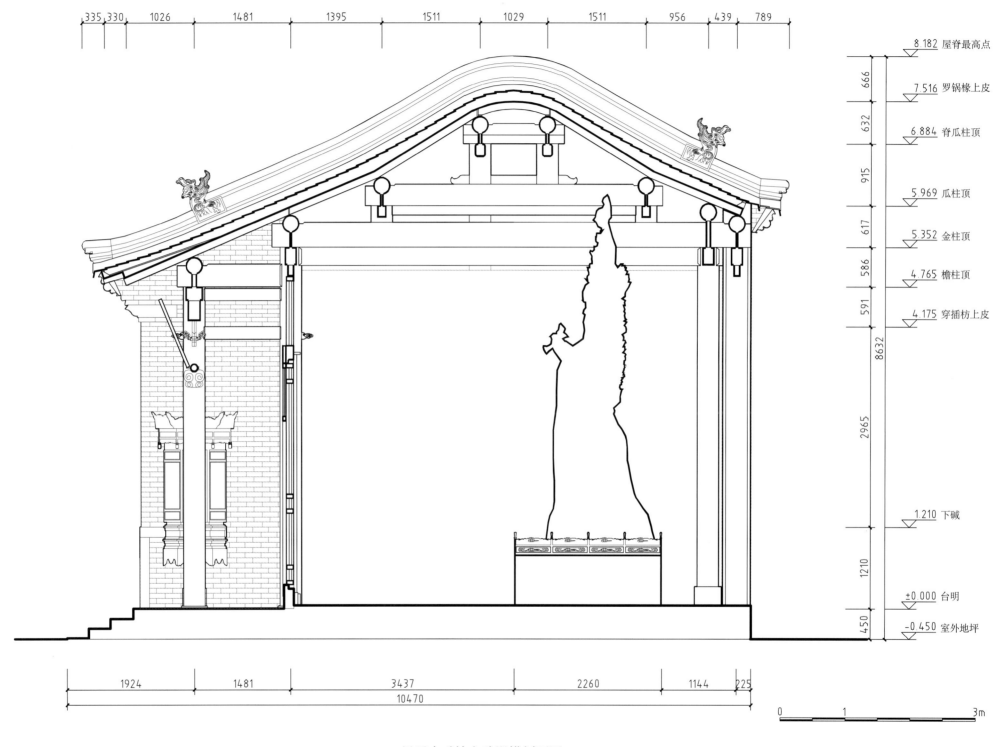

335 330　1026　　1481　　1395　　1511　　1029　　1511　　956　439　789

8.182 屋脊最高点
666
7.516 罗锅椽上皮
632
6.884 脊瓜柱顶
915
5.969 瓜柱顶
617
5.352 金柱顶
586
4.765 檐柱顶
591
4.175 穿插枋上皮
8632
2965
1.210 下碱
1210
±0.000 台明
450
-0.450 室外地坪

1924　　1481　　　3437　　　2260　　1144　225
10470

0　　1　　3m

显通寺千钵文殊殿横剖面图
Cross-section of Qianbo wenshudian of Xiantongsi

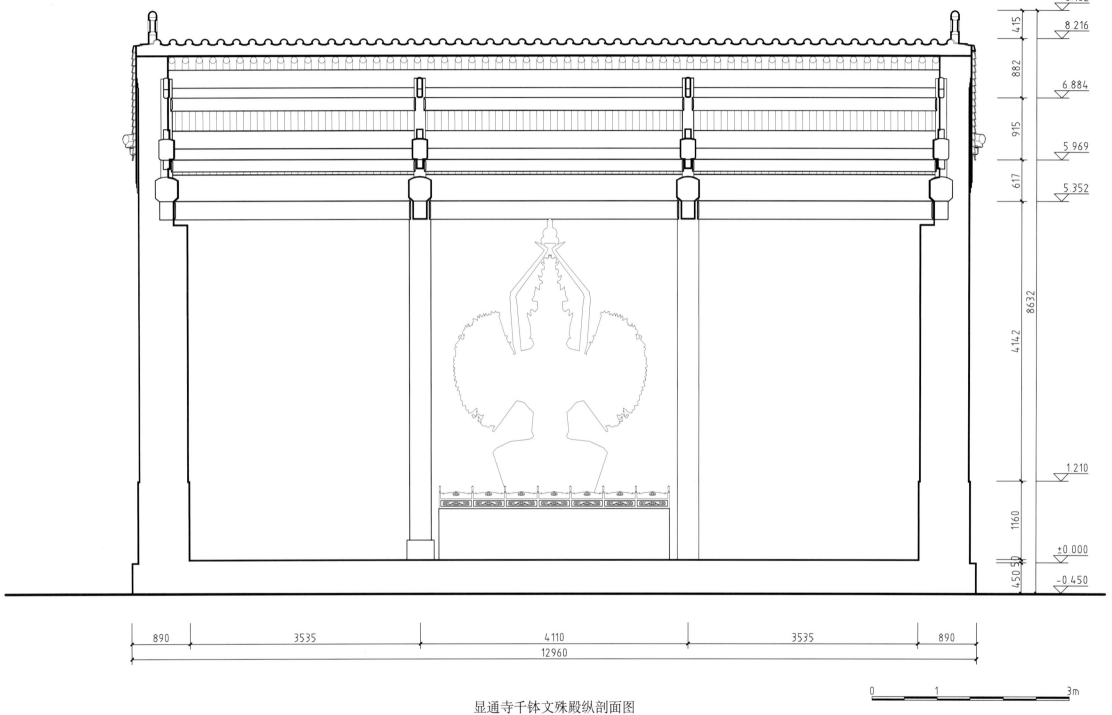

8.182

8.216

415

6.884

882

5.969

915

5.352

617

8632

1.210

4142

1160

±0.000

450

-0.450

890 3535 4110 3535 890

12960

显通寺千钵文殊殿纵剖面图

Longitudinal section of Qianbo wenshudian of Xiantongsi

0 1 3m

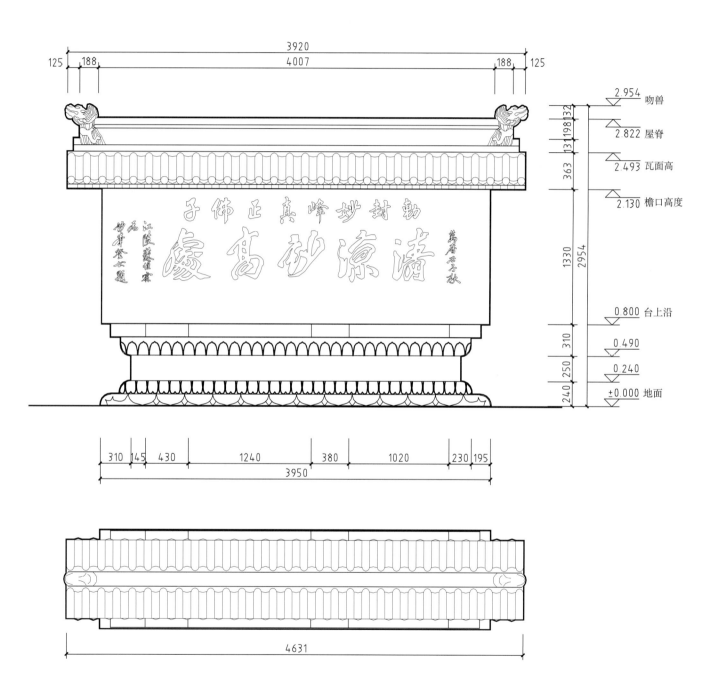

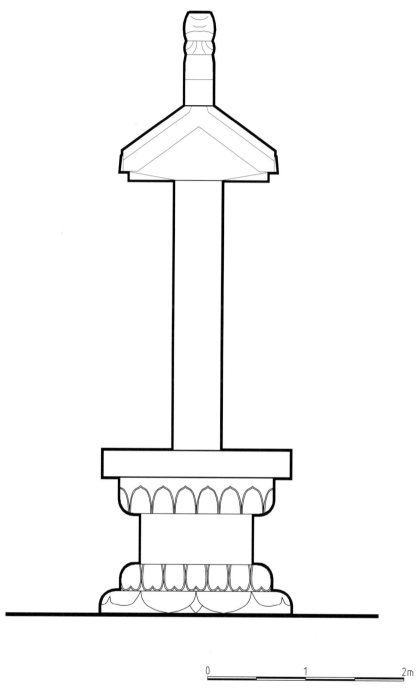

显通寺影壁正立面图及屋顶平面图
Front elevation and roof plan of *yingbi* of Xiantongsi

显通寺影壁侧立面图
Side elevation of *yingbi* of Xiantongsi

显通寺塔幢平面及立面图

Plan and elevation of *tachuang* of Xiantongsi

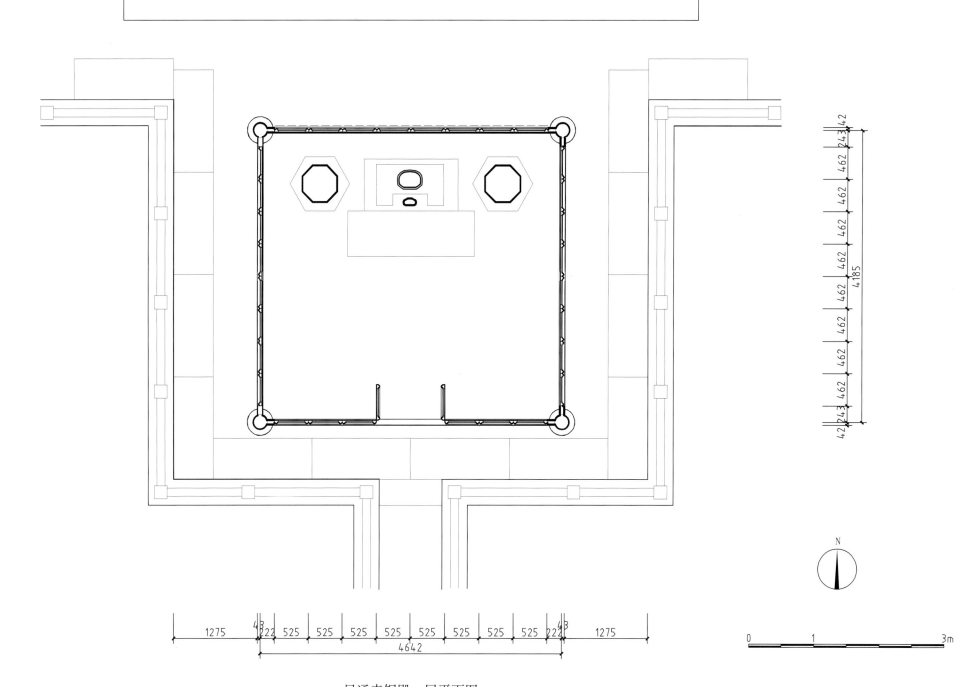

显通寺铜殿一层平面图

Plan of first floor of *tongdian* of Xiantongsi

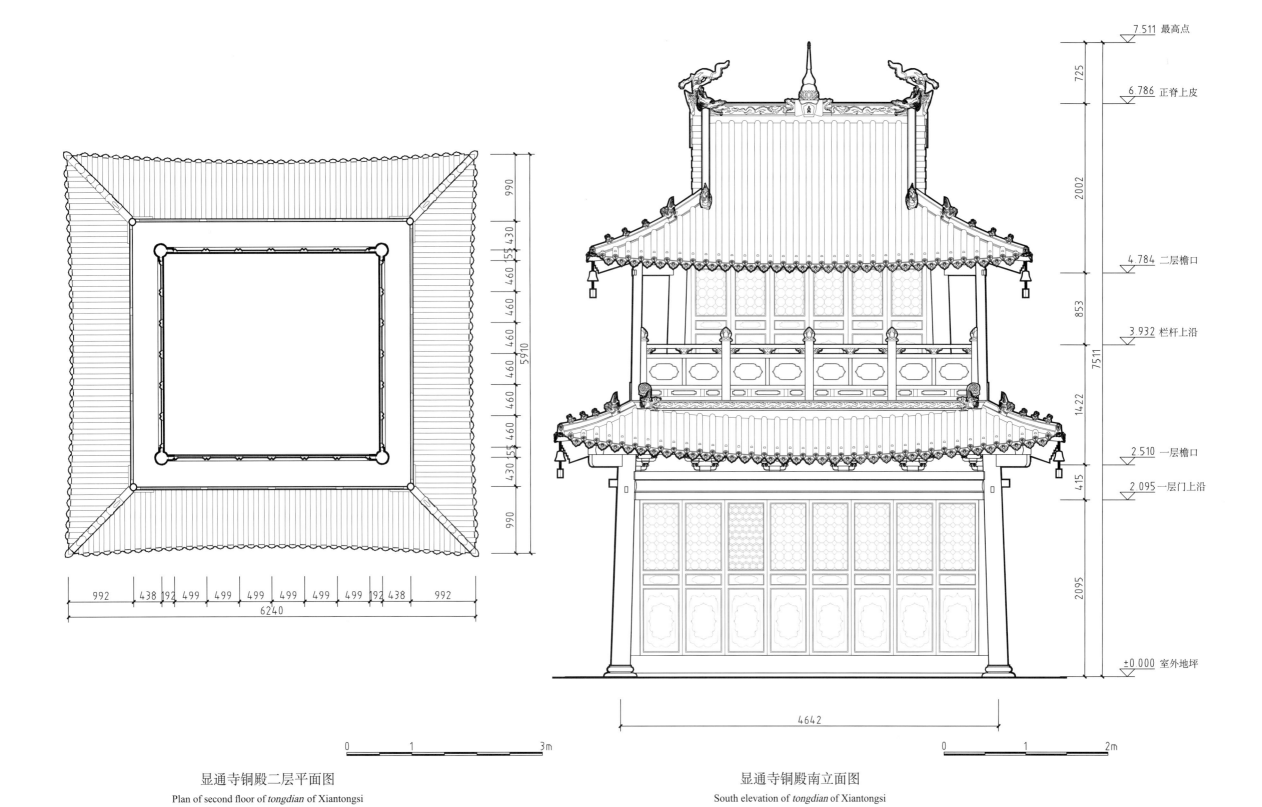

7 511 最高点

6 786 正脊上皮

4 784 二层檐口

3 932 栏杆上沿

2 510 一层檐口

2 095 一层门上沿

±0 000 室外地坪

725

2002

853

1422

415

2095

7511

6240

5910

990

155 430

460

460

460

460

155 430

990

992 438 192 499 499 499 499 499 499 192 438 992

4642

0   1   3m

0   1   2m

显通寺铜殿二层平面图

Plan of second floor of *tongdian* of Xiantongsi

显通寺铜殿南立面图

South elevation of *tongdian* of Xiantongsi

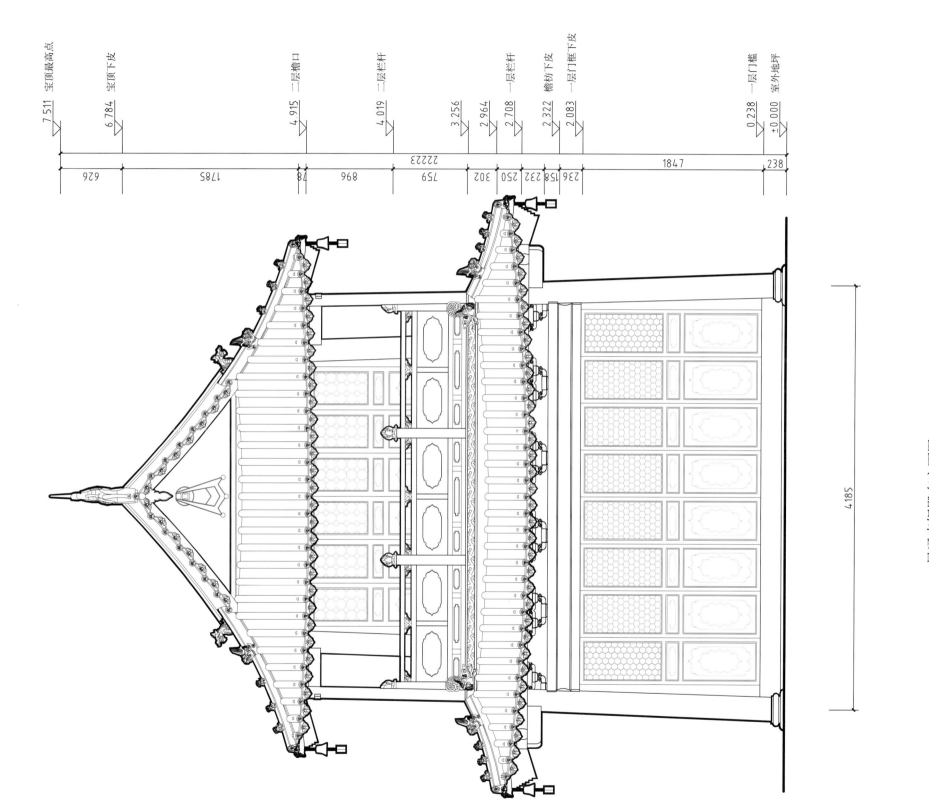

宝顶最高点 7 511
宝顶下皮 6 784
二层檐口 4 915
二层栏杆 4 019
一层栏杆 3 256
2 964
檐枋下皮 2 708
一层门框下皮 2 322
2 083
一层门槛 0 238
室外地坪 ±0 000

626
1785
78
896
759
302 250 232 158 236
1847
238
22223

4 185

2m
1
0

显通寺铜殿东立面图
East elevation of *tongdian* of Xiantongsi

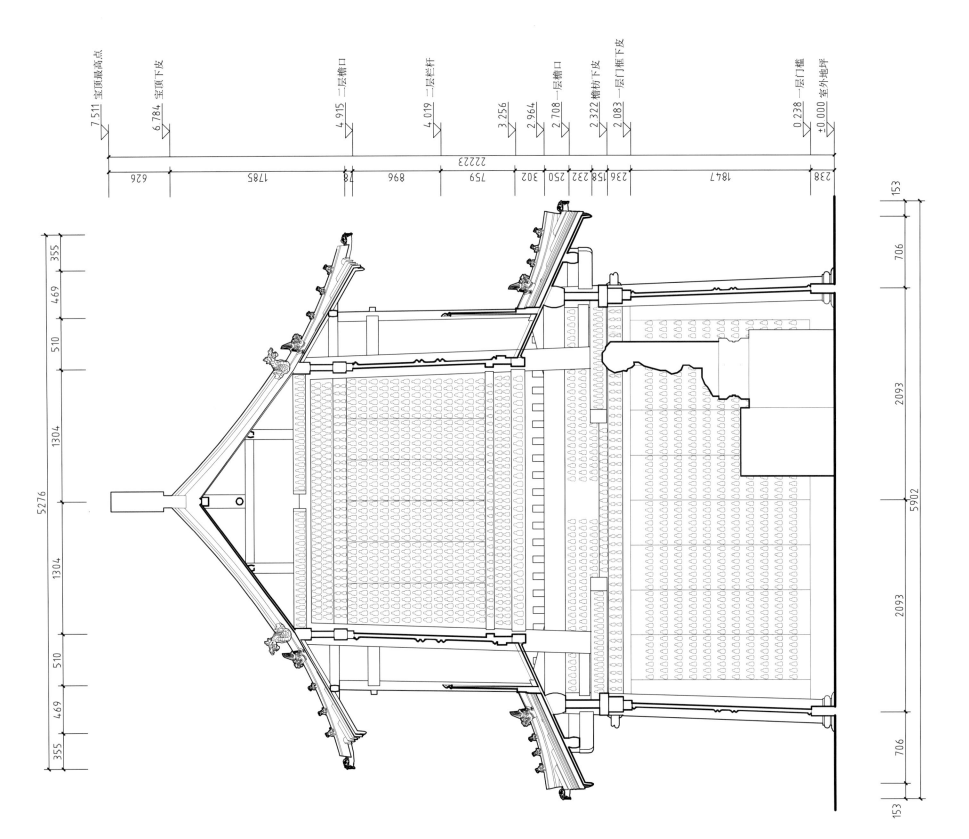

7 511 宝顶最高点
6 784 宝顶下皮
4 915 二层檐口
4 019 二层栏杆
3 256
2 964
2 708 一层檐口
2 322 檐枋下皮
2 083 一层门框下皮
一层门槛
0 238
±0.000 室外地坪

显通寺铜殿横剖面图
Cross-section of *tongdian* of Xiantongsi

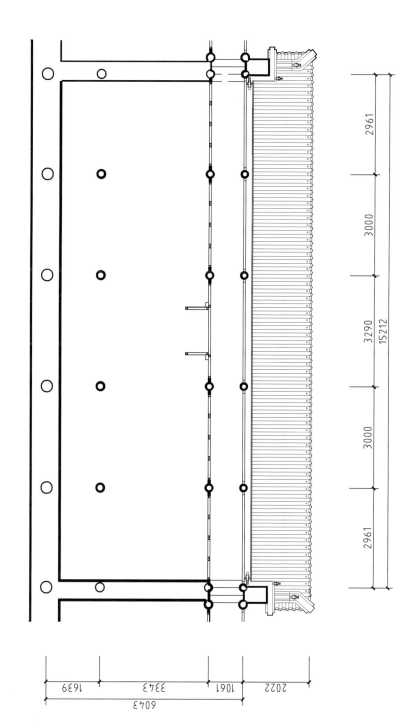

显通寺藏经殿一层平面图
Plan of first floor of *cangjingdian* of Xiantongsi

显通寺藏经殿二层平面图
Plan of second floor of *cangjingdian* of Xiantongsi

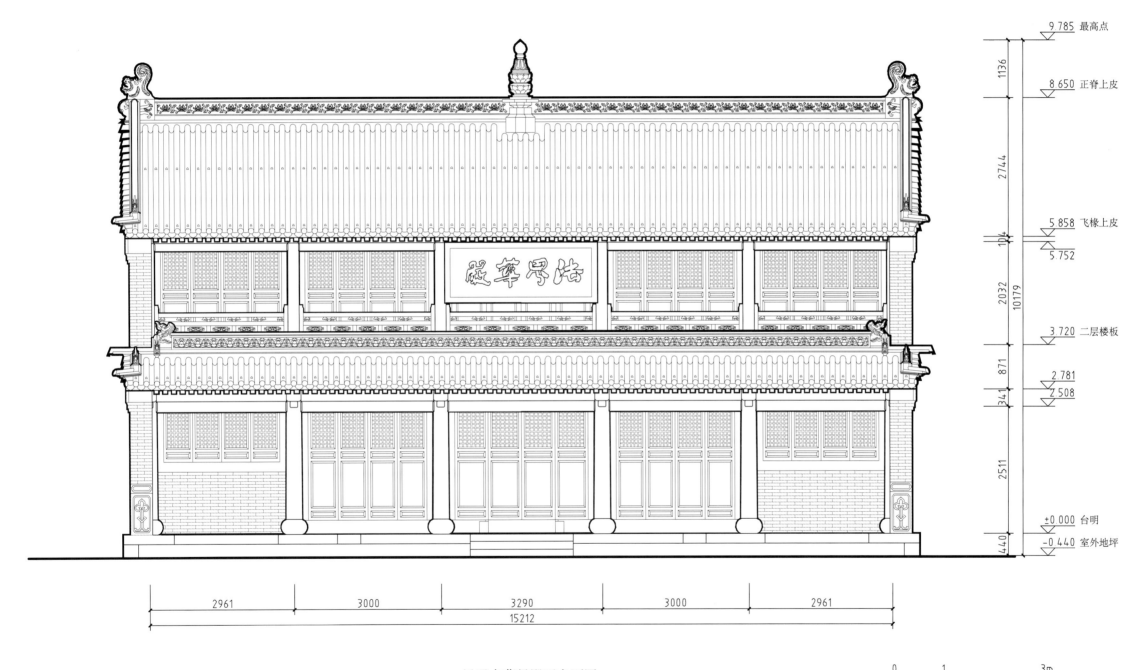

9.785 最高点

8.650 正脊上皮

5.858 飞椽上皮

5.752

3.720 二层楼板

2.781

2.508

±0.000 台明

-0.440 室外地坪

1136

2744

104

2032

10179

871

341

2511

440

2961   3000   3290   3000   2961

15212

显通寺藏经殿正立面图

Front elevation of *cangjingdian* of Xiantongsi

0   1   3m

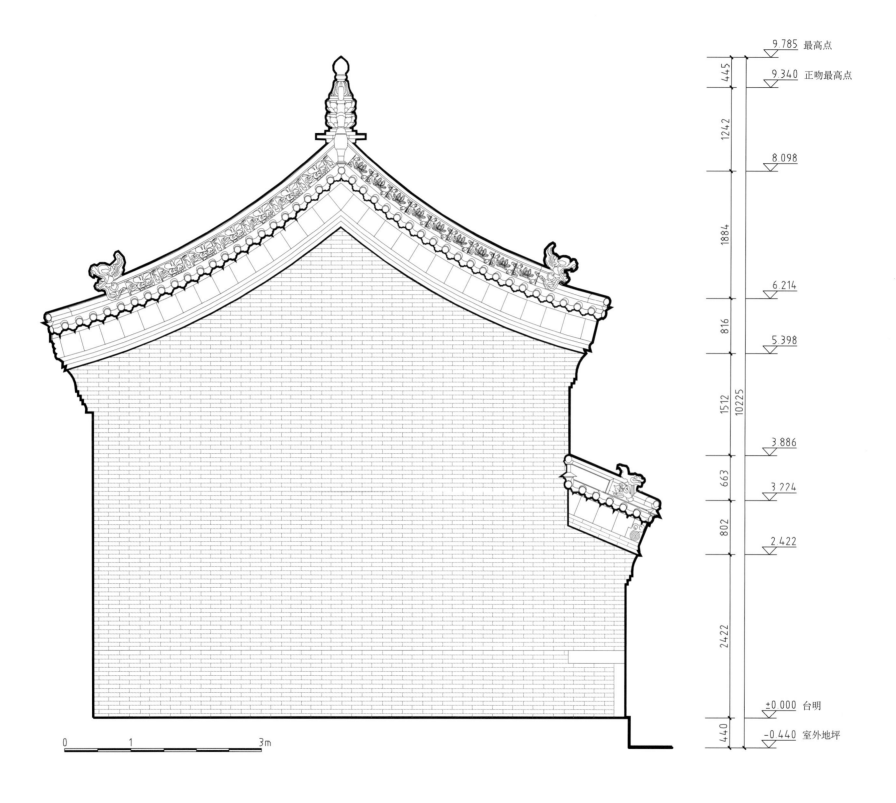

9.785 最高点

445

9.340 正吻最高点

1242

8.098

1884

6.214

816

5.398

1512

10225

3.886

663

3.224

802

2.422

2422

±0.000 台明

440

-0.440 室外地坪

0　　1　　3m

显通寺藏经殿侧立面图

Side elevation of *cangjingdian* of Xiantongsi

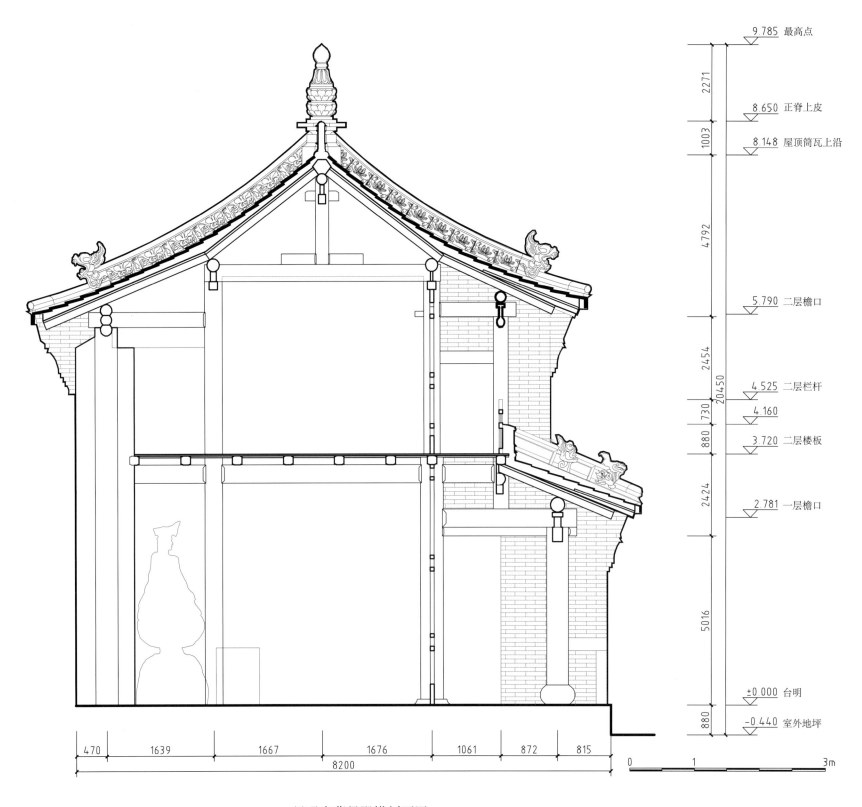

9 785 最高点

8 650 正脊上皮

8 148 屋顶筒瓦上沿

5 790 二层檐口

4 525 二层栏杆

4 160

3 720 二层楼板

2 781 一层檐口

±0.000 台明

-0.440 室外地坪

2271

1003

4792

2454

730

880

2424

5016

20450

880

470　1639　1667　1676　1061　872　815

8200

0　1　　　　3m

显通寺藏经殿横剖面图

Cross-section of *cangjingdian* of Xiantongsi

073

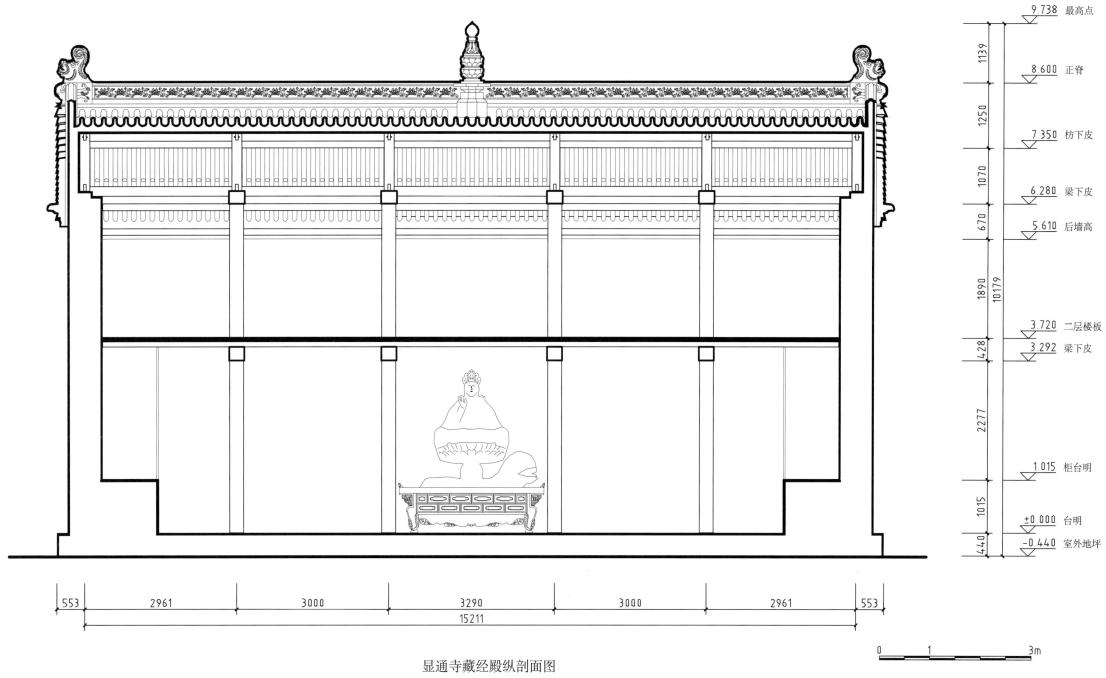

9.738 最高点
1139
8.600 正脊
1250
7.350 枋下皮
1070
6.280 梁下皮
670
5.610 后墙高
1890
10179
3.720 二层楼板
428
3.292 梁下皮
2277
1.015 柜台明
1015
±0.000 台明
440
-0.440 室外地坪

553 2961 3000 3290 3000 2961 553
15211

0   1   3m

显通寺藏经殿纵剖面图
Longitudinal section of *cangjingdian* of Xiantongsi

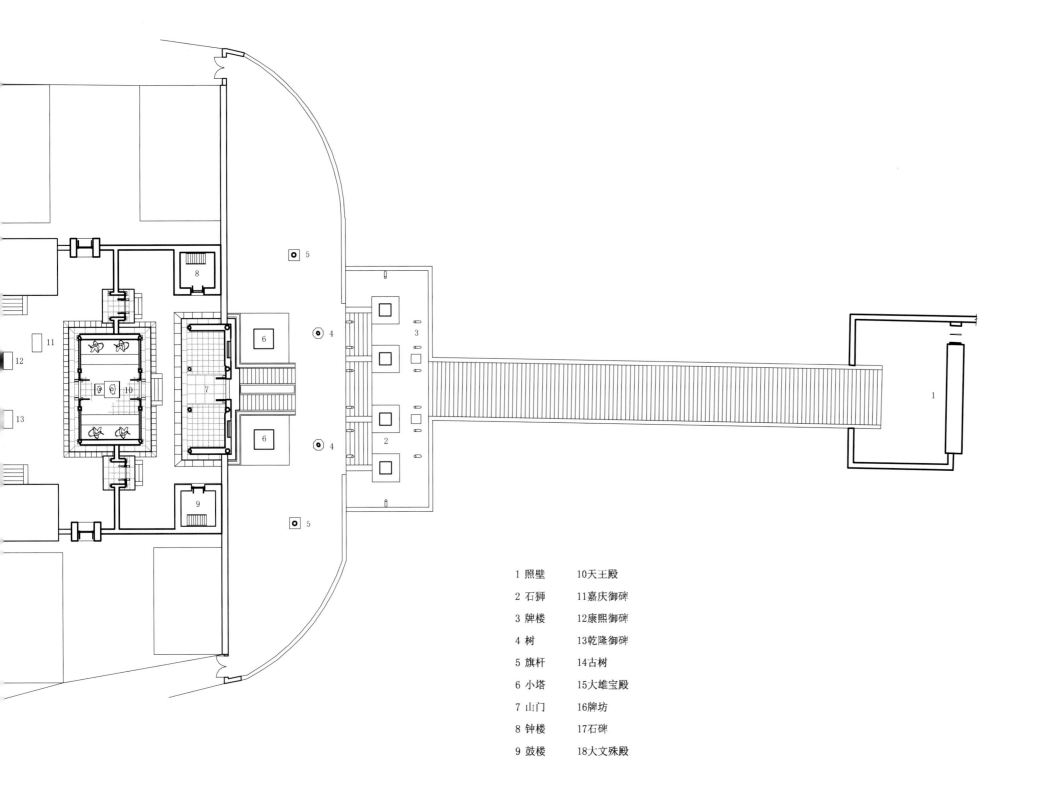

中国古建筑测绘大系·宗教建筑——五台山佛教建筑

1 照壁　　　10天王殿
2 石狮　　　11嘉庆御碑
3 牌楼　　　12康熙御碑
4 树　　　　13乾隆御碑
5 旗杆　　　14古树
6 小塔　　　15大雄宝殿
7 山门　　　16牌坊
8 钟楼　　　17石碑
9 鼓楼　　　18大文殊殿

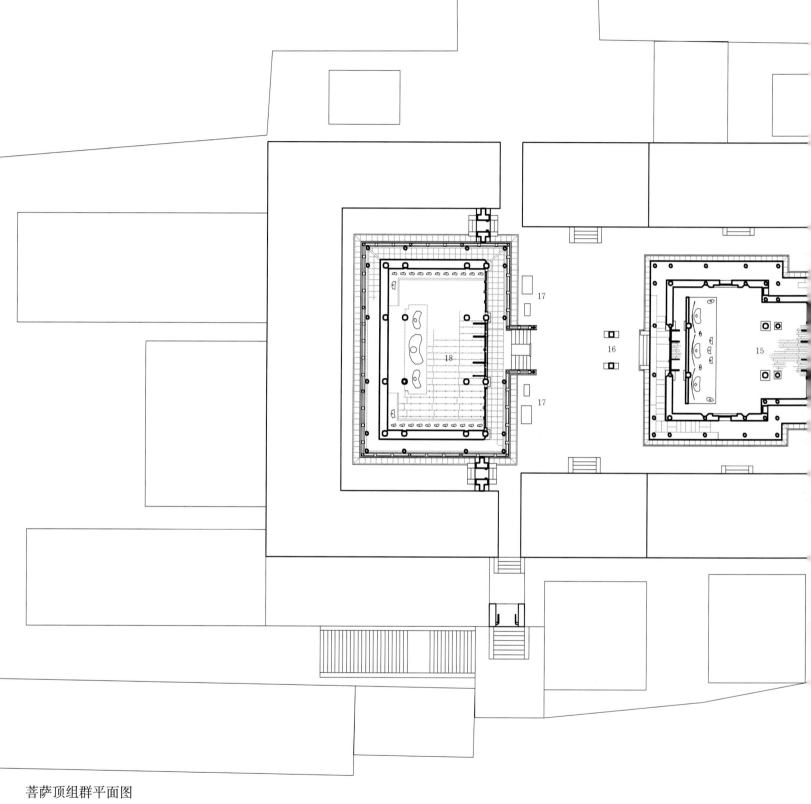

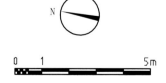

0　1　　　　　5m

N

菩萨顶组群平面图
Grop plan of Bodhisattva Summit

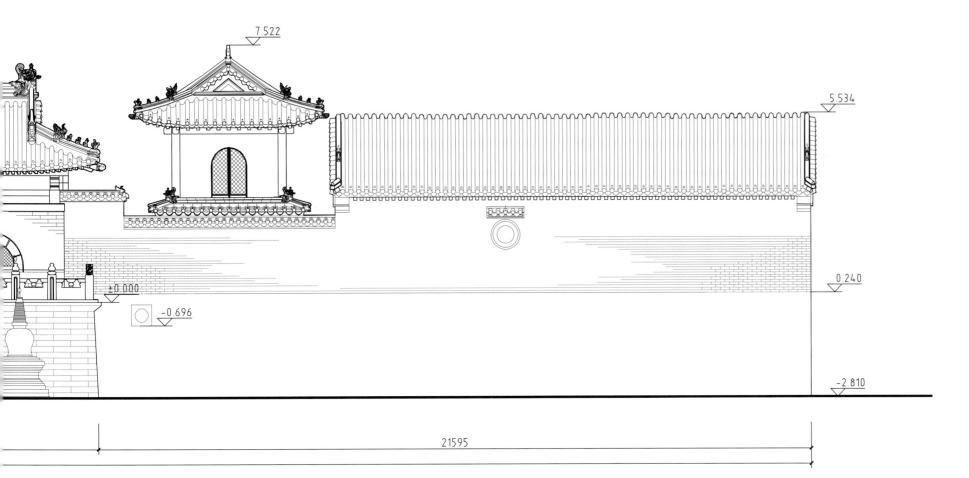

7.522

5.534

0.240

±0.000

-0.696

-2.810

21595

0   1   3m

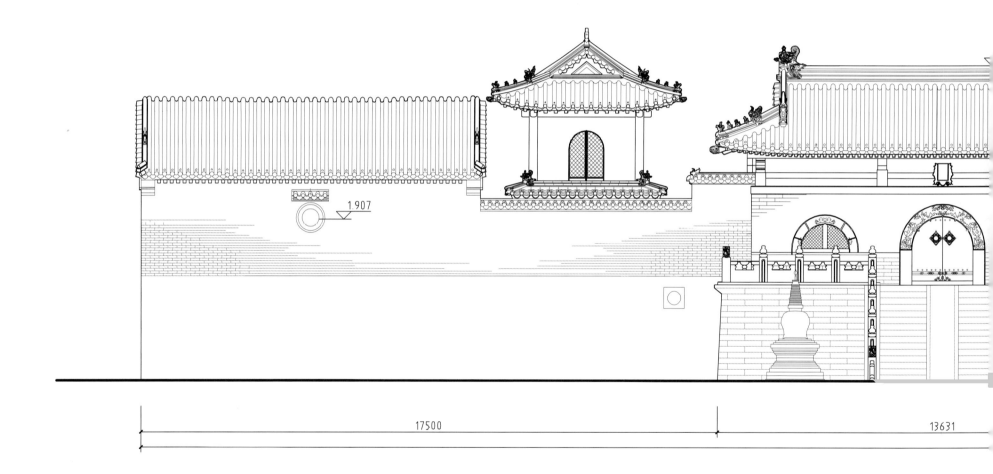

1.907

17500   13631

菩萨顶入口处正立面图

Front elevation of entrance of Bodhisattva Summit

9 055

11700   2780   5420   8455   6729

0 1   3m

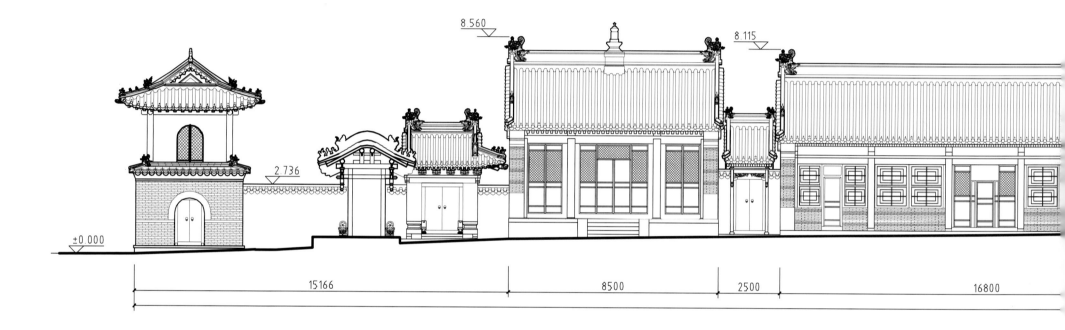

菩萨顶院内立面图
Elevation of courtyard of Bodhisattva Summit

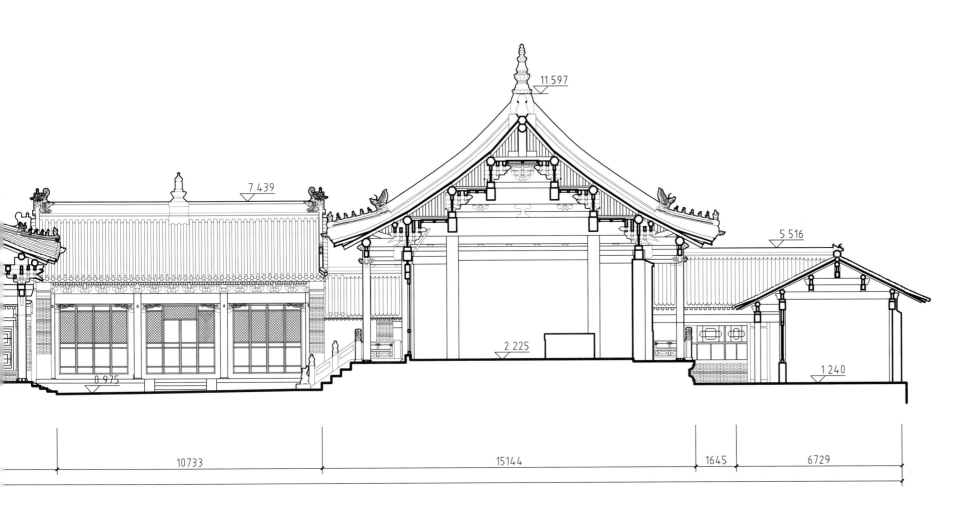

11.597

7.439

5.516

2.225

1.240

0.975

10733

15144

1645

6729

0 1 3m

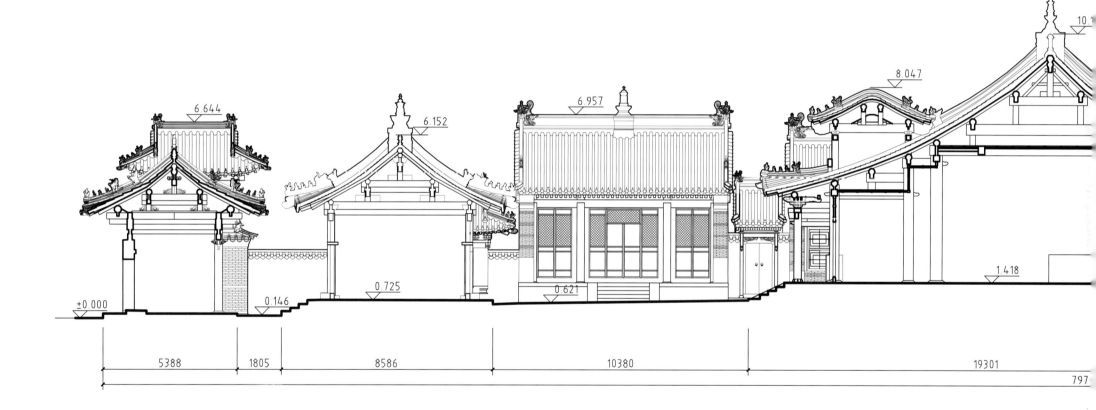

6.644

6.152

6.957

8.047

10

+0.000

0.146

0.725

0.621

1.418

5388   1805   8586   10380   19301

797

菩萨顶总剖面图

Site section of Bodhisattva Summit

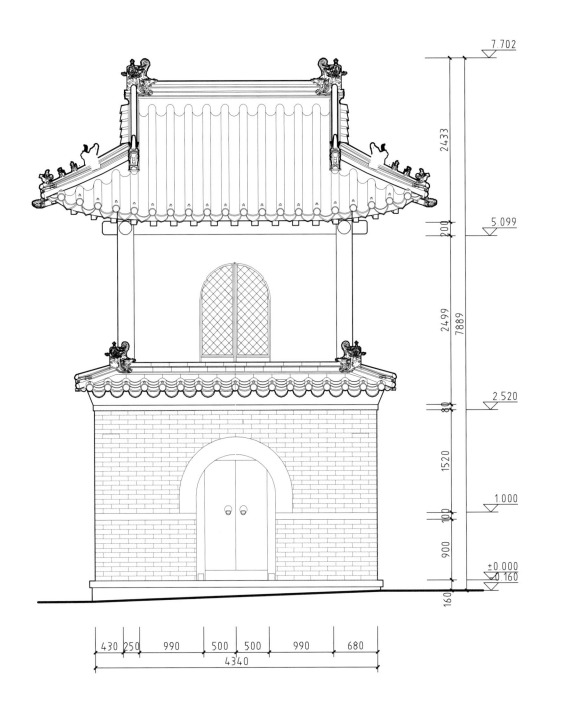

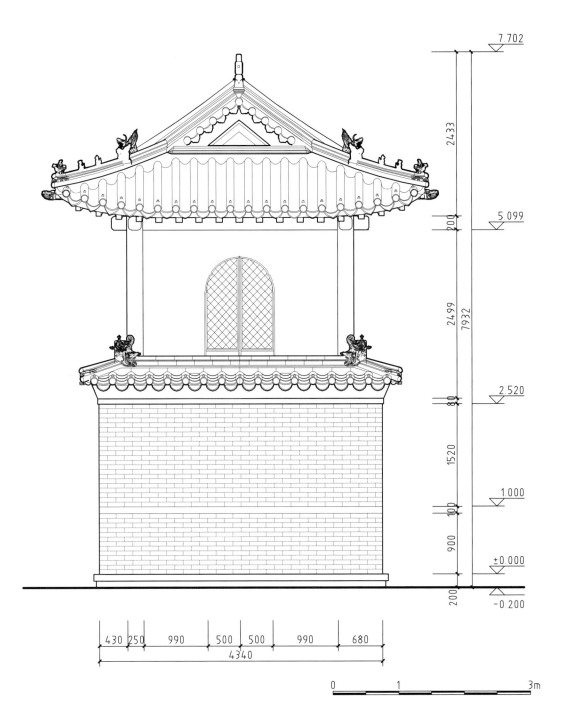

菩萨顶钟鼓楼立面图

Elevation of *zhonglou* and *gulou* of Bodhisattva Summit

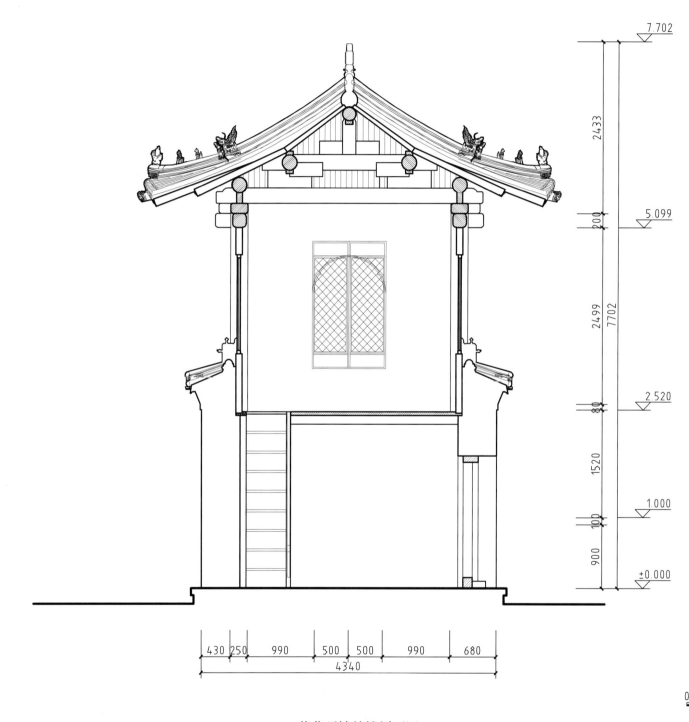

菩萨顶钟鼓楼剖面图

Section of *zhonglou* and *gulou* of Bodhisattva Summit

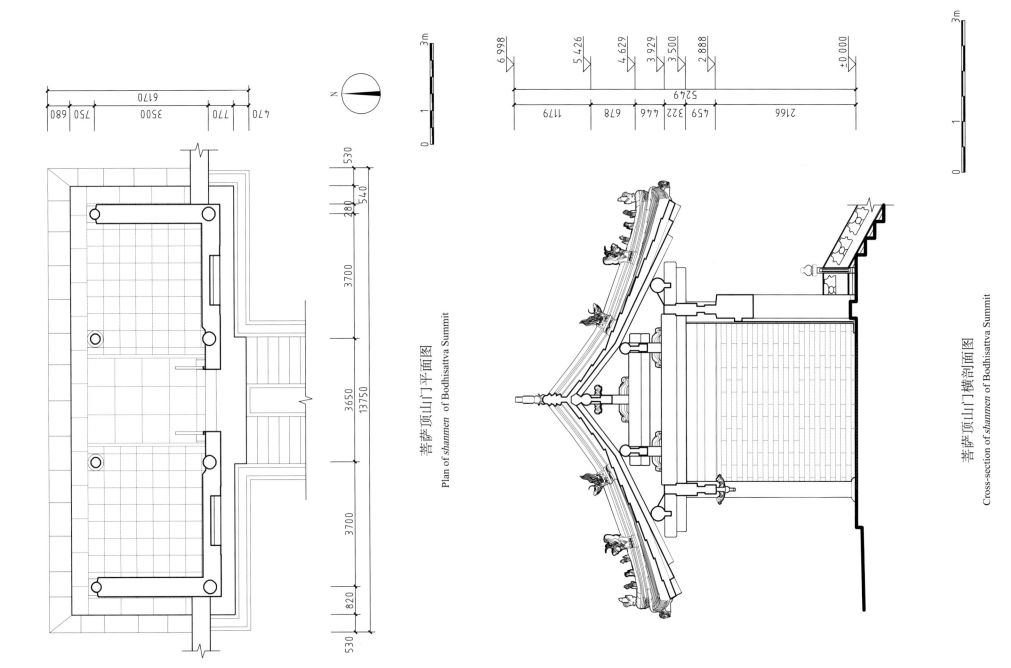

菩萨顶山门平面图
Plan of *shanmen* of Bodhisattva Summit

菩萨顶山门横剖面图
Cross-section of *shanmen* of Bodhisattva Summit

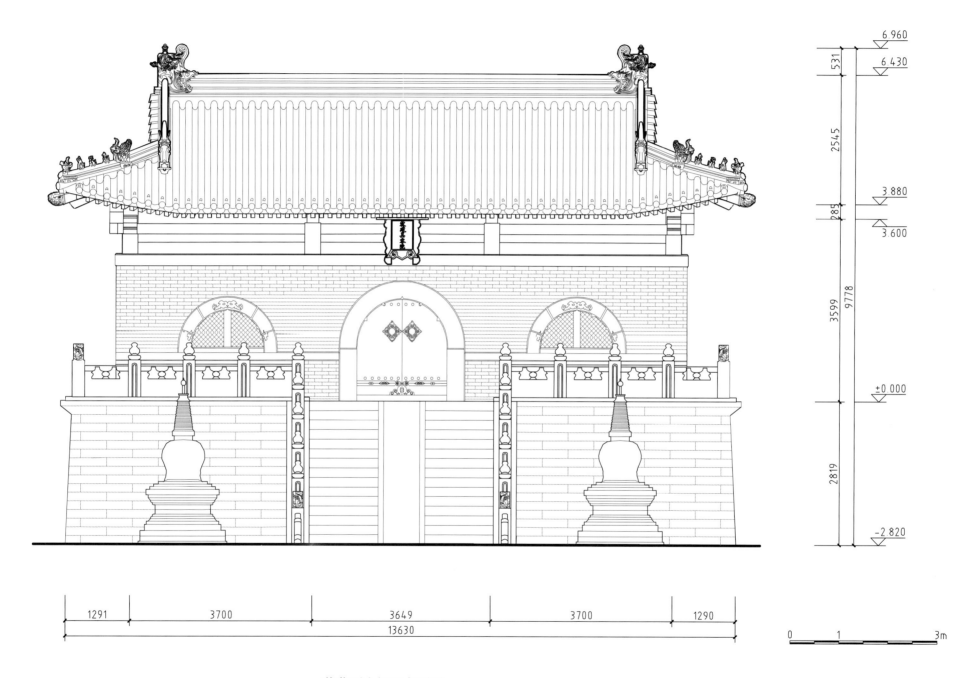

菩萨顶山门正立面图

Front elevation of *shanmen* of Bodhisattva Summit

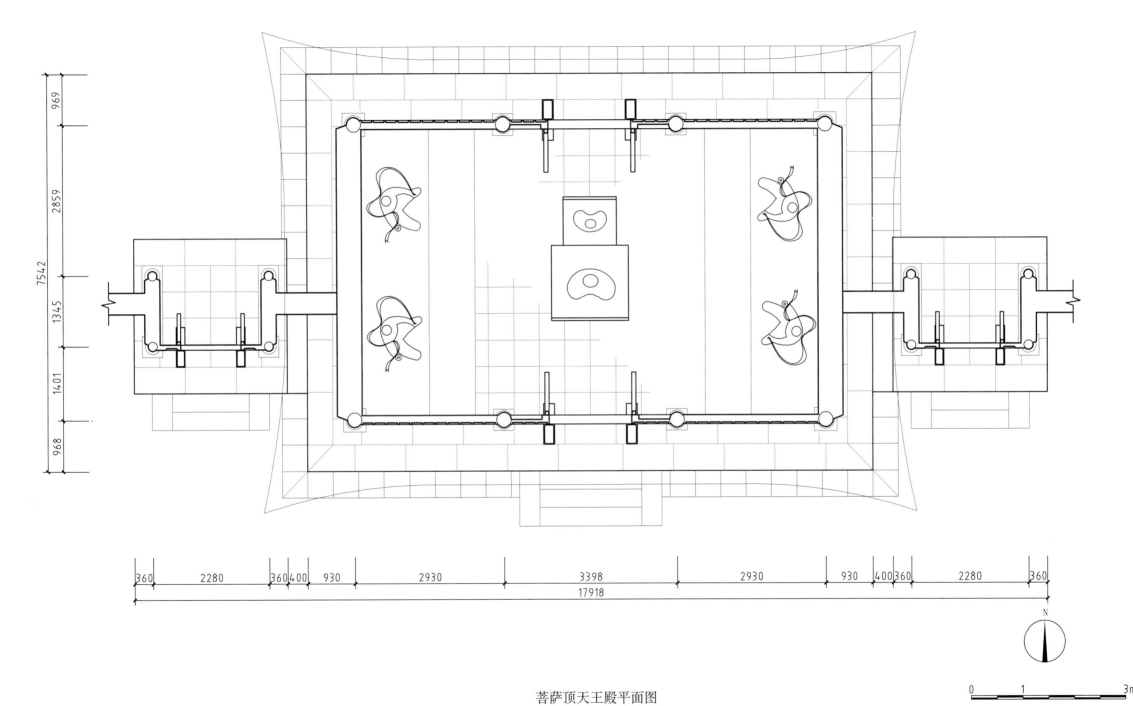

菩萨顶天王殿平面图
Plan of Tianwangdian of Bodhisattva Summit

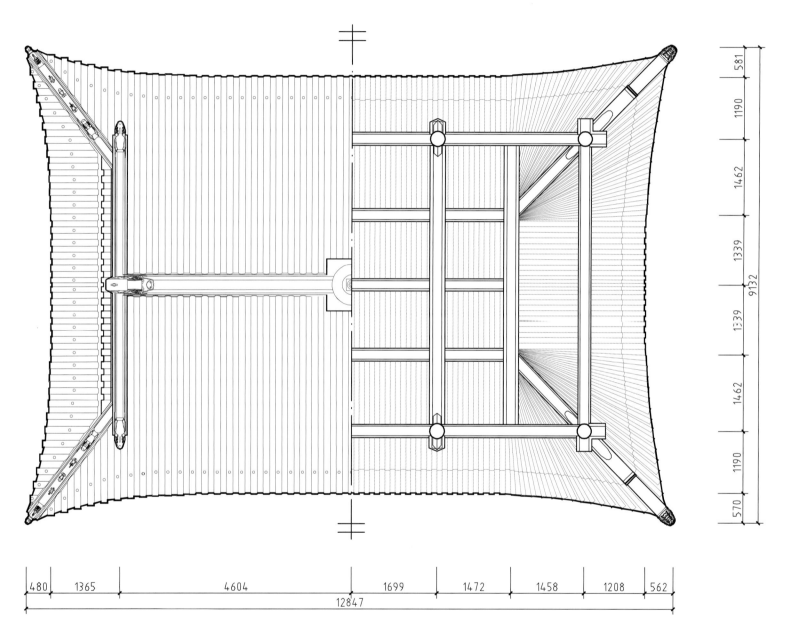

581
1190
1462
1339
9132
1:39
1462
1190
570

480 1365 4604 1699 1472 1458 1208 562
12847

菩萨顶天王殿屋顶平面图
Roof plan of Tianwangdian of Bodhisattva Summit

0 1 3m

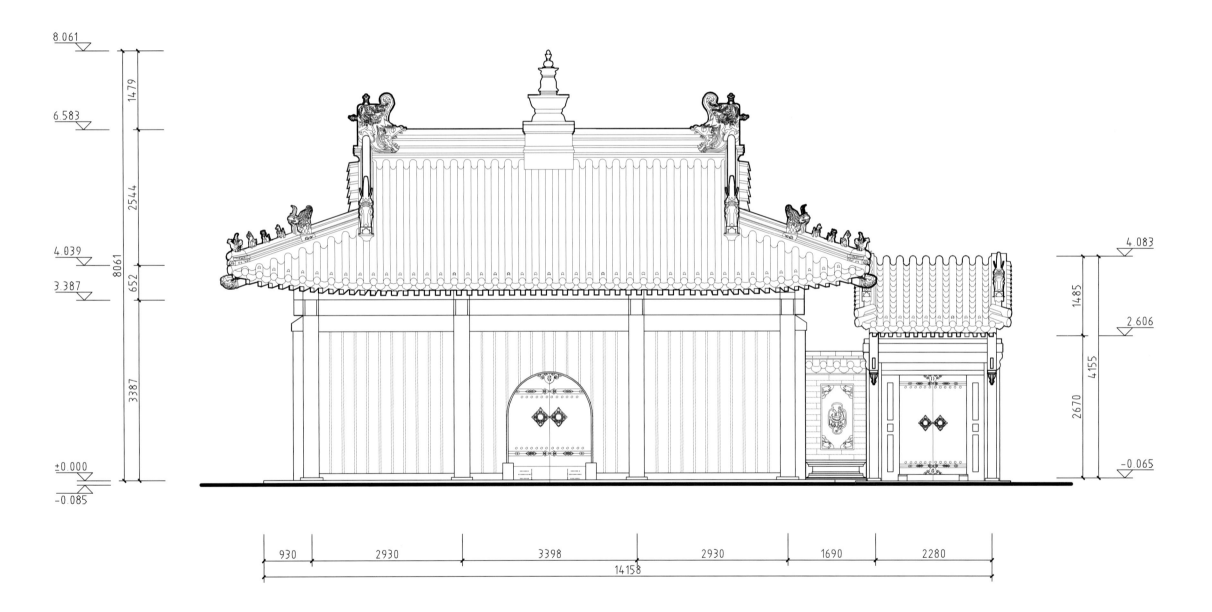

8.061

1479

6.583

2544

4.039

8061

652

3.387

3387

±0.000

-0.085

4.083

1485

2.606

4155

2670

-0.065

930　2930　3398　2930　1690　2280

14158

菩萨顶天王殿南立面图
South elevation of Tianwangdian of Bodhisattva Summit

0　1　2m

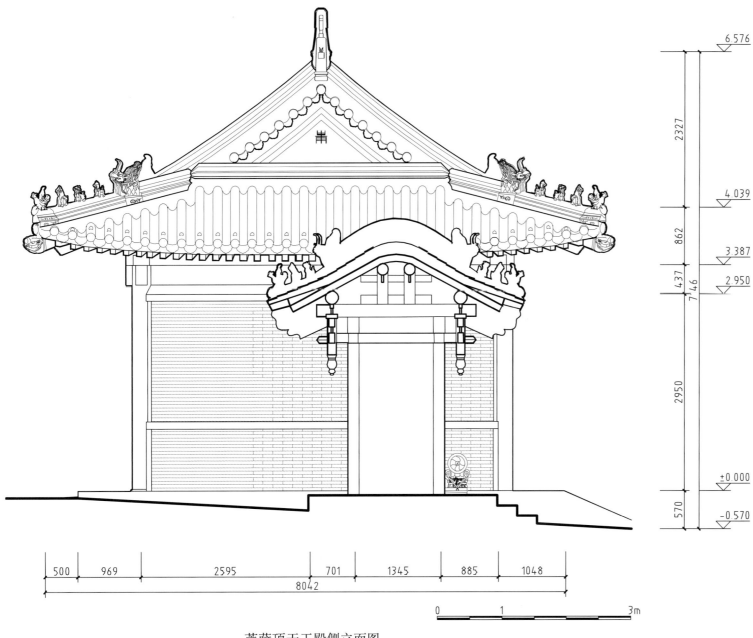

6 576

2327

4 039

862

3 387

4 37

7 46

2 950

2950

± 0 000

570

-0 570

500 969 2595 701 1345 885 1048

8042

0   1   3m

310

250

630

380

360

0   0.1   0 3m

菩萨顶天王殿侧立面图

Side elevation of Tianwangdian of Bodhisattva Summit

菩萨顶天王殿抱鼓石大样图

Baogushi of Tianwangdian of Bodhisattva Summit

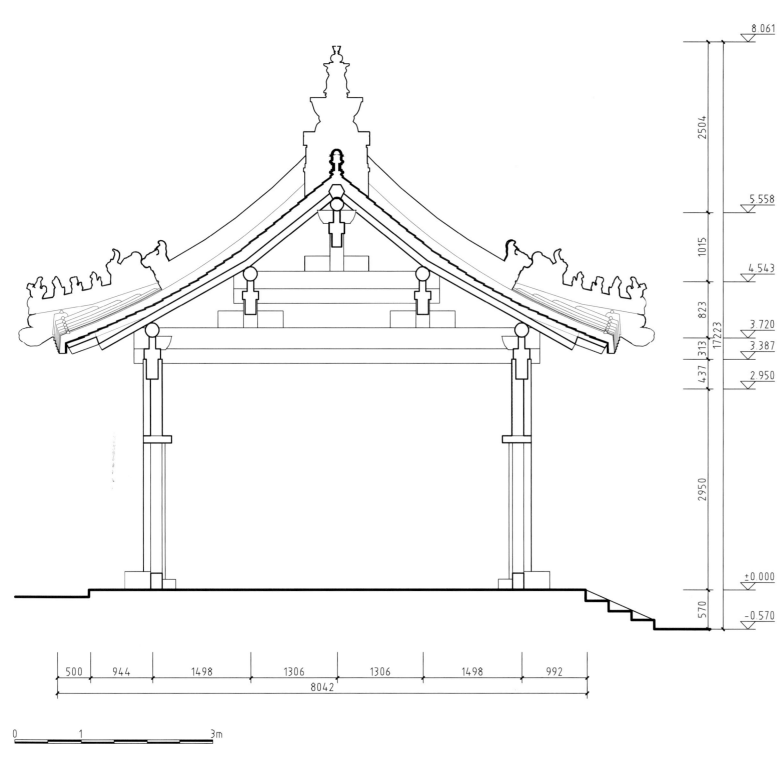

菩萨顶天王殿横剖面图
Cross-section of Tianwangdian of Bodhisattva Summit

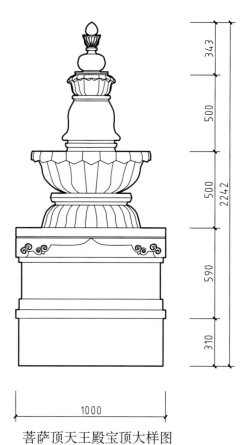

菩萨顶天王殿宝顶大样图
Roof Plan of Tianwangdian of Bodhisattva Summit

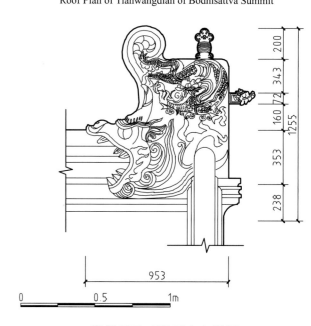

菩萨顶天王殿正吻大样图
Zhengwen of Tianwangdian of Bodhisattva Summit

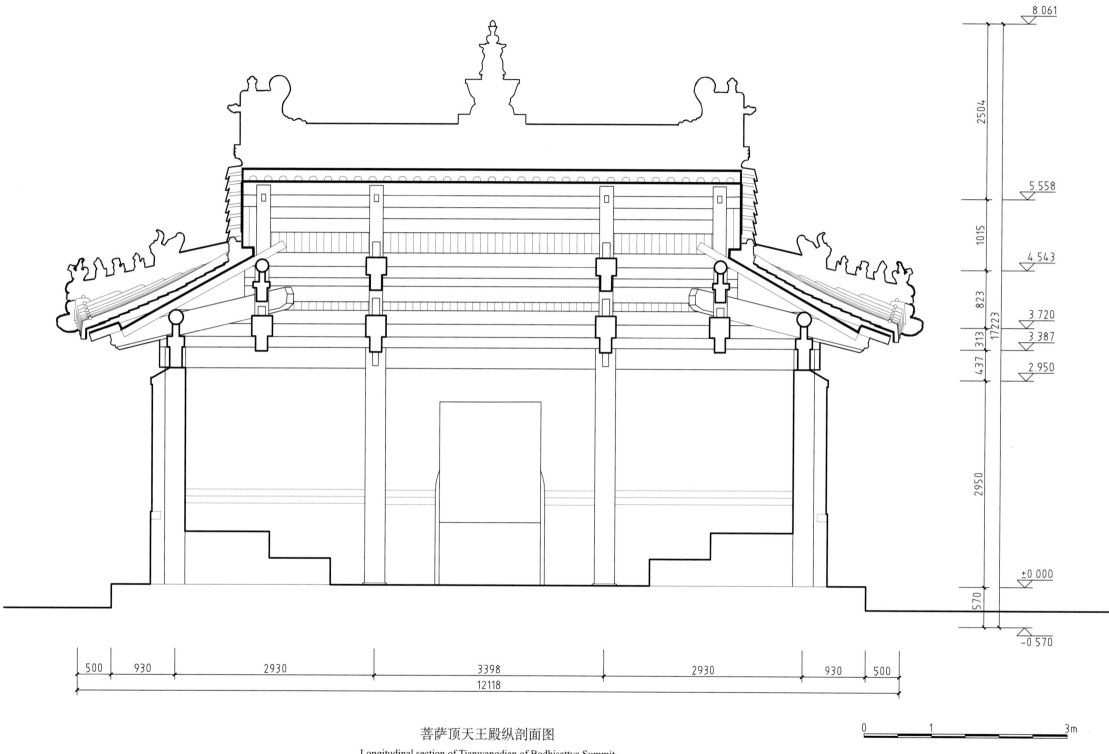

菩萨顶天王殿纵剖面图
Longitudinal section of Tianwangdian of Bodhisattva Summit

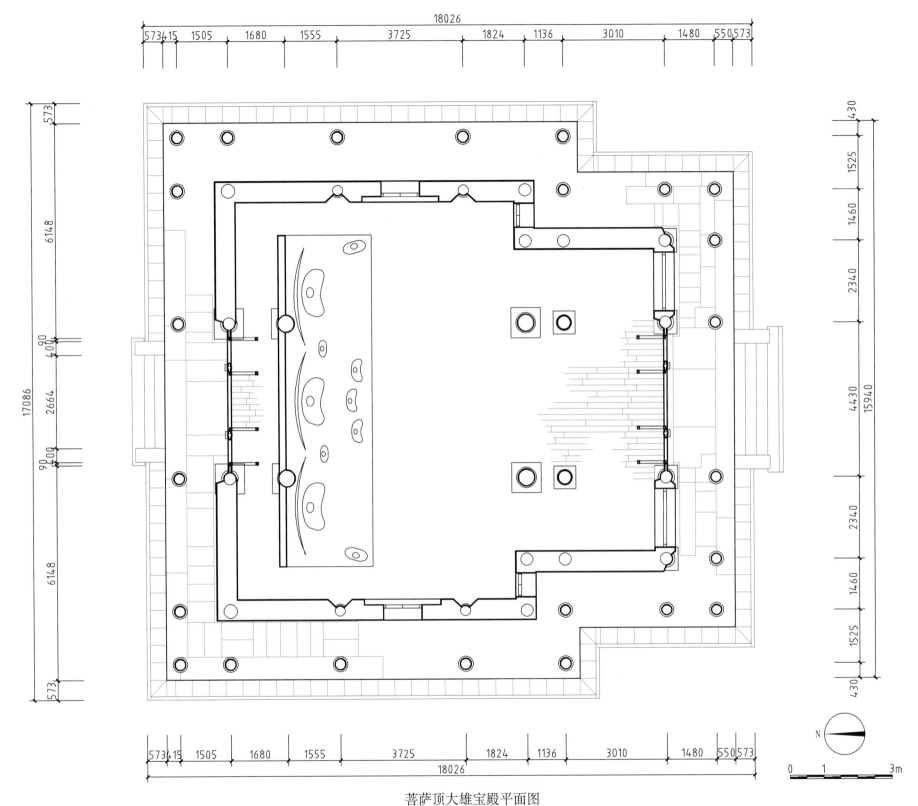

菩萨顶大雄宝殿平面图
Plan of Daxiong baodian of Bodhisattva Summit

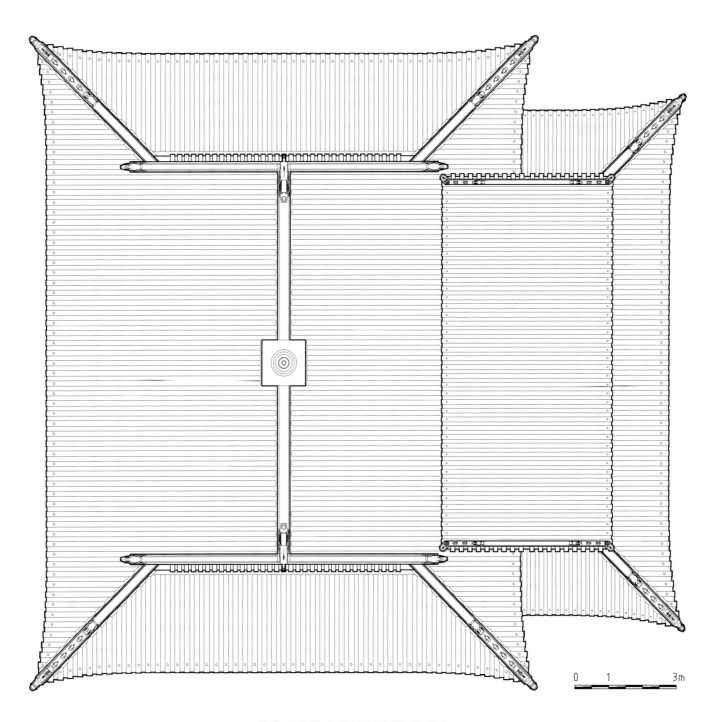

菩萨顶大雄宝殿屋顶俯视平面图

Plan of roof of Daxiong baodian as seen from above of Bodhisattva Summit

0　　1　　　　3m

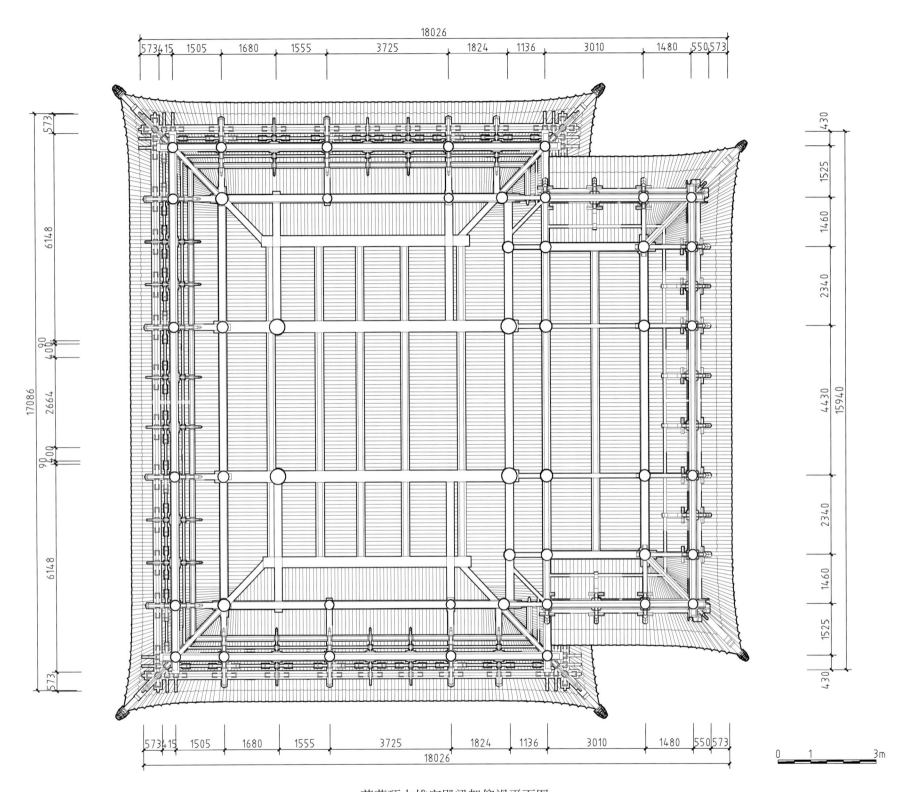

菩萨顶大雄宝殿梁架仰视平面图

Plan of roof of Daxiong baodian of Bodhisattva Summit as seen from below

0 1 3m

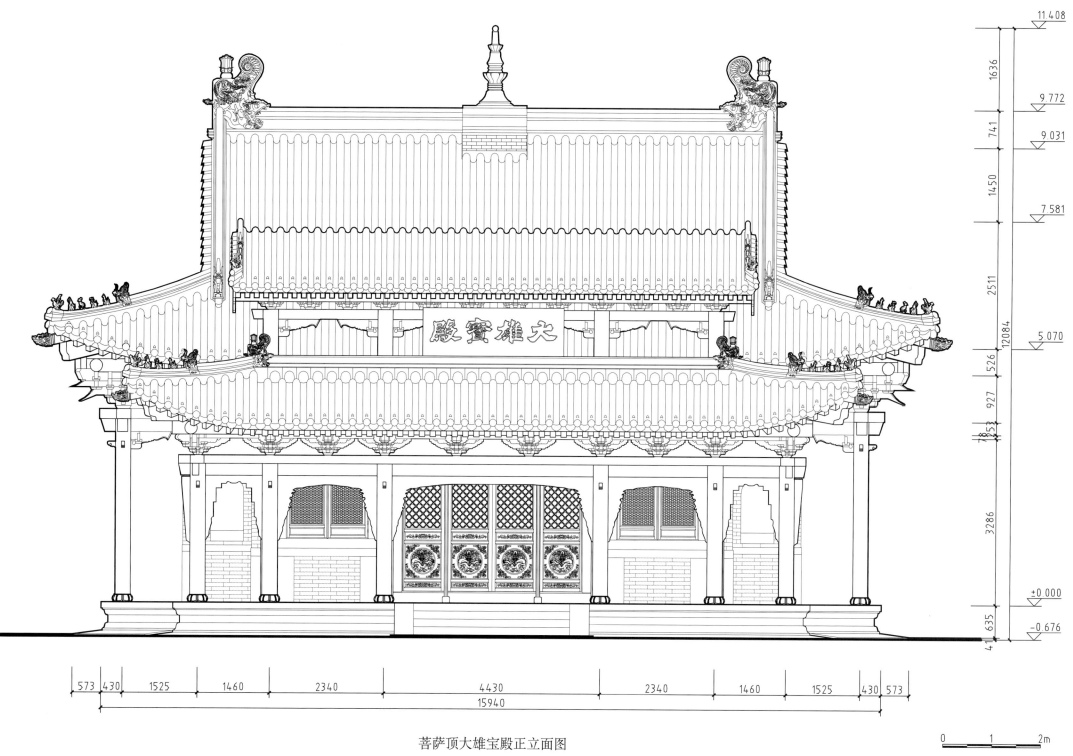

11.408
1636
9.772
741
9.031
1450
7.581
2511
12084
5.070
526
927
753
3286
+0.000
635
-0.676

573 430 1525 1460 2340 4430 2340 1460 1525 430 573
15940

大雄寶殿

菩萨顶大雄宝殿正立面图
Front elevation of Daxiong baodian of Bodhisattva Summit

0 1 2m

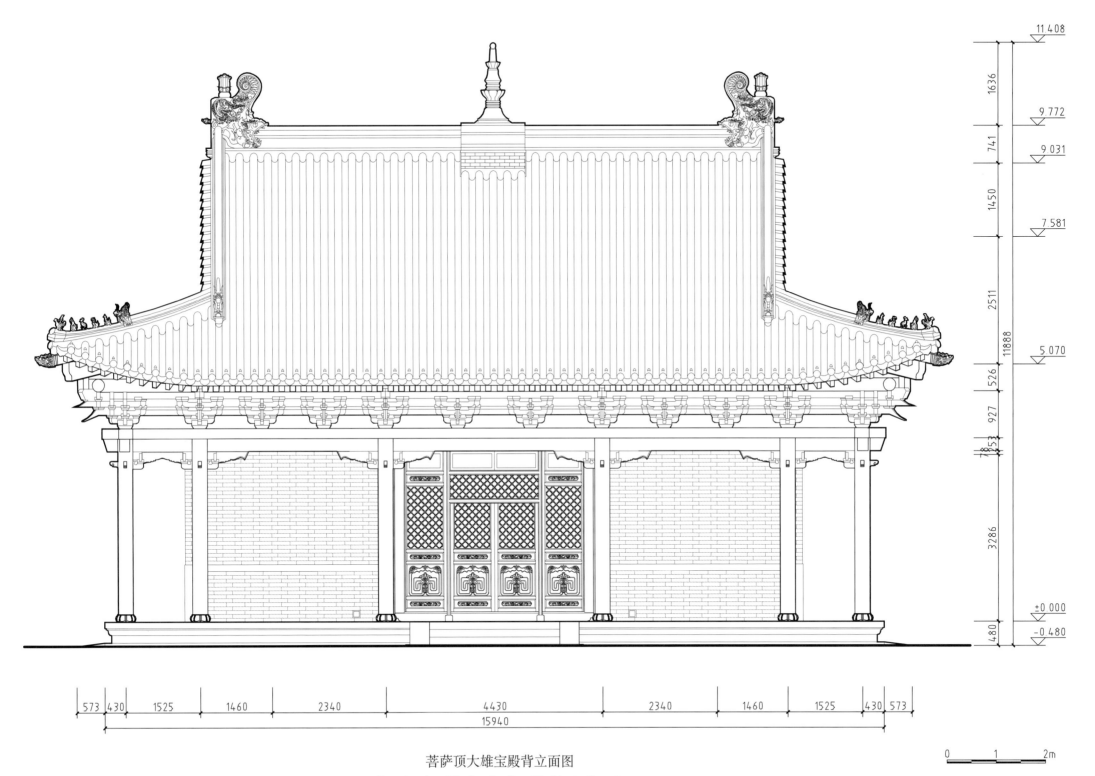

11 408

9 772

9 031

7 581

5 070

±0 000

-0 480

1636

741

1450

2511

526

927

853

3286

480

11888

573 430 1525 1460 2340 4430 2340 1460 1525 430 573

15940

菩萨顶大雄宝殿背立面图

Rear elevation of Daxiong baodian of Bodhisattva Summit

0 1 2m

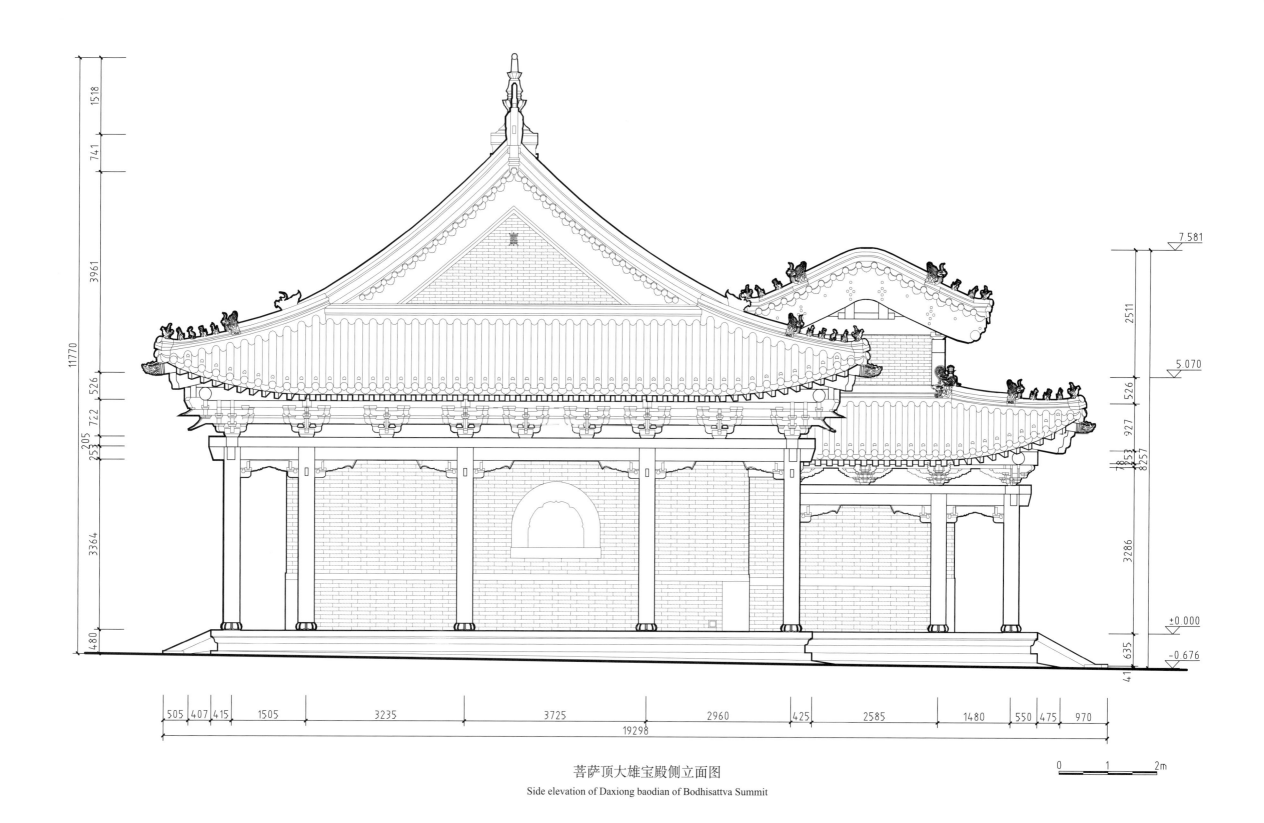

菩萨顶大雄宝殿侧立面图

Side elevation of Daxiong baodian of Bodhisattva Summit

0   1   2m

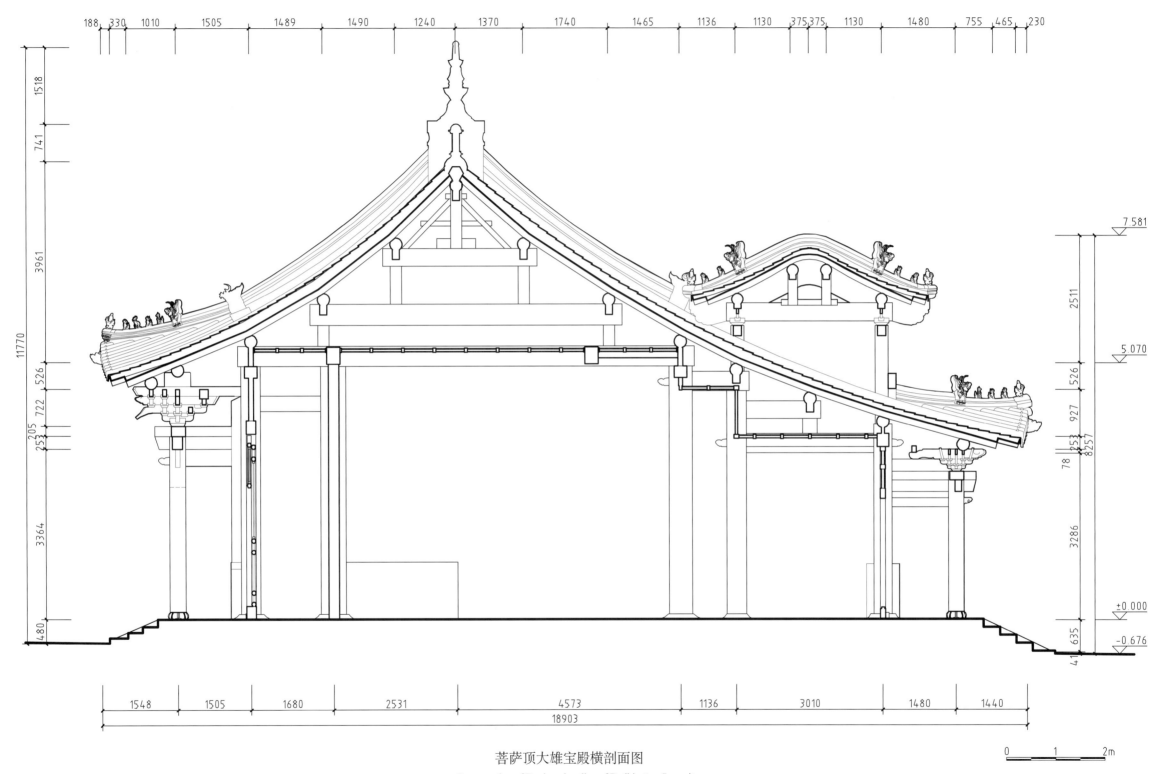

菩萨顶大雄宝殿横剖面图

Cross-section of Daxiong baodian of Bodhisattva Summit

0   1   2m

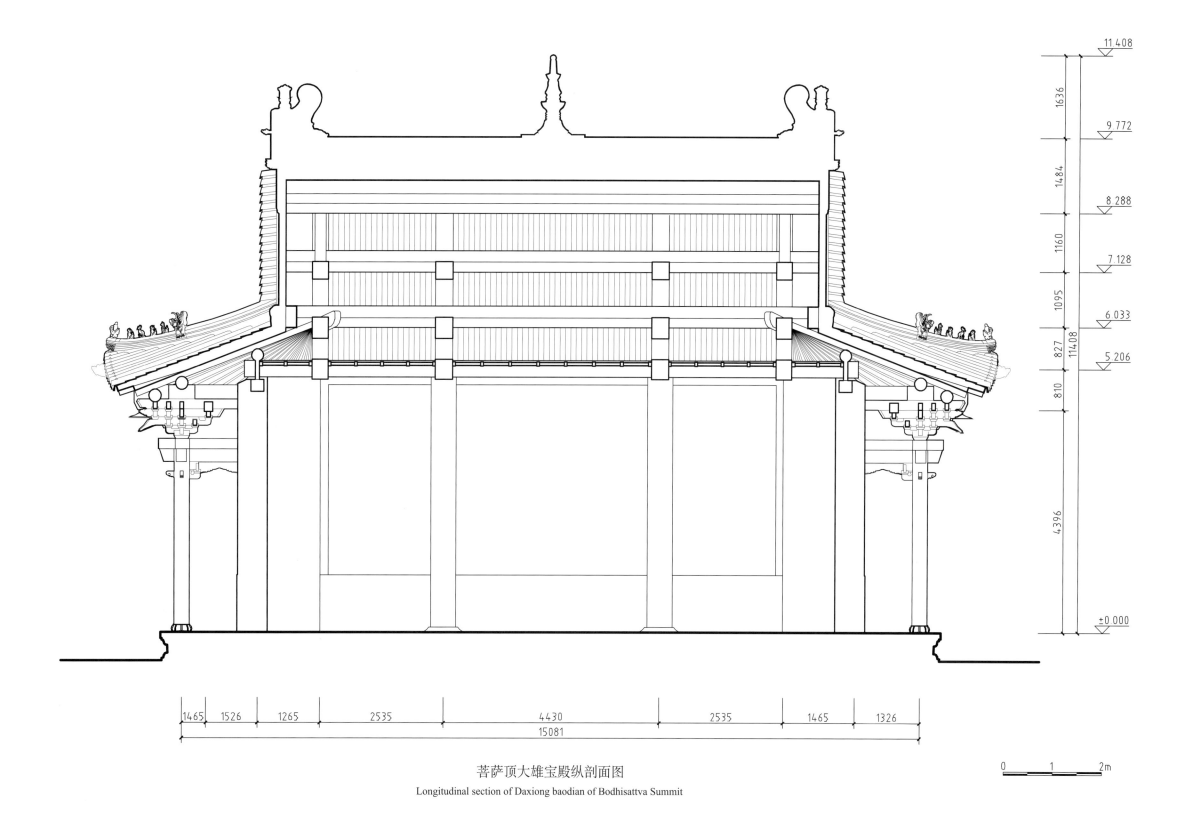

11.408

1636

9.772

1484

8.288

1160

7.128

1095

6.033

827

11408

5.206

810

4396

±0.000

1465 | 1526 | 1265 | 2535 | 4430 | 2535 | 1465 | 1326

15081

菩萨顶大雄宝殿纵剖面图

Longitudinal section of Daxiong baodian of Bodhisattva Summit

0   1   2m

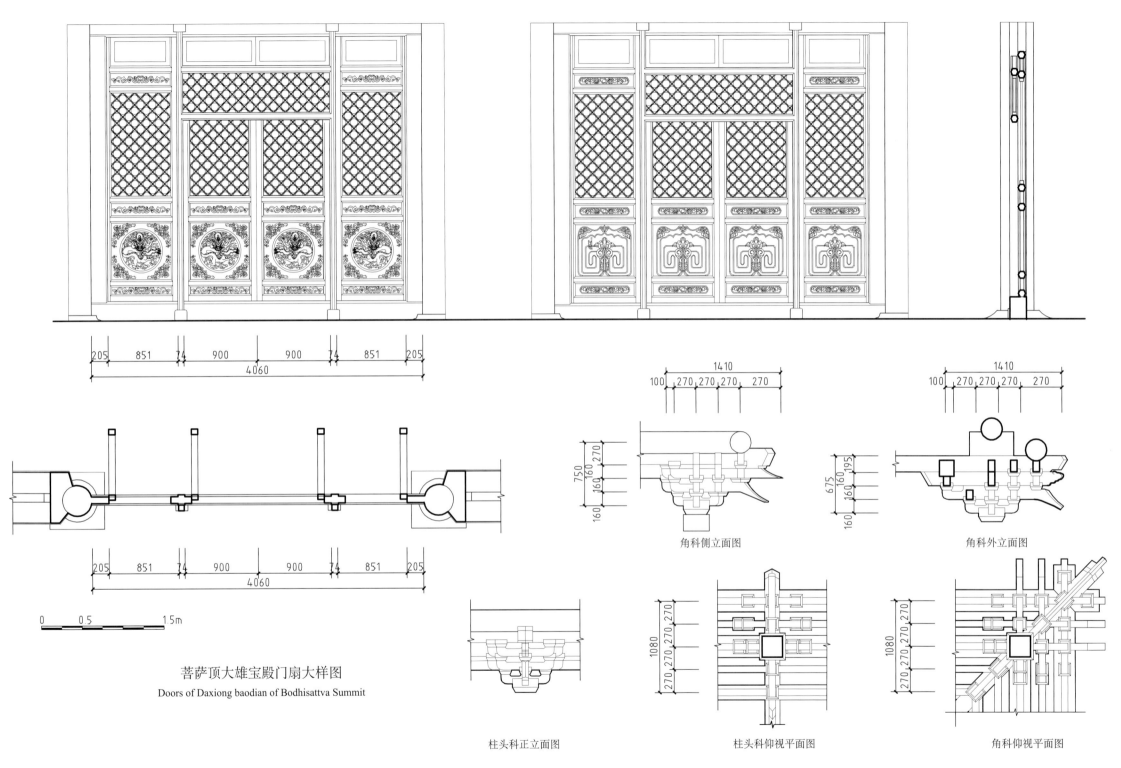

205 851 74 900 900 74 851 205
4060

205 851 74 900 900 74 851 205
4060

0　0.5　　　1.5m

菩萨顶大雄宝殿门扇大样图
Doors of Daxiong baodian of Bodhisattva Summit

1410
100 270 270 270 270

750
160 270
160
160

角科侧立面图

1410
100 270 270 270 270

675
160 195
160
160

角科外立面图

1080
270 270 270
270

柱头科正立面图

1080
270 270 270
270 270 270

柱头科仰视平面图

角科仰视平面图

菩萨顶大雄宝殿斗栱大样图
Bracketing of Daxiong baodian of Bodhisattva Summit

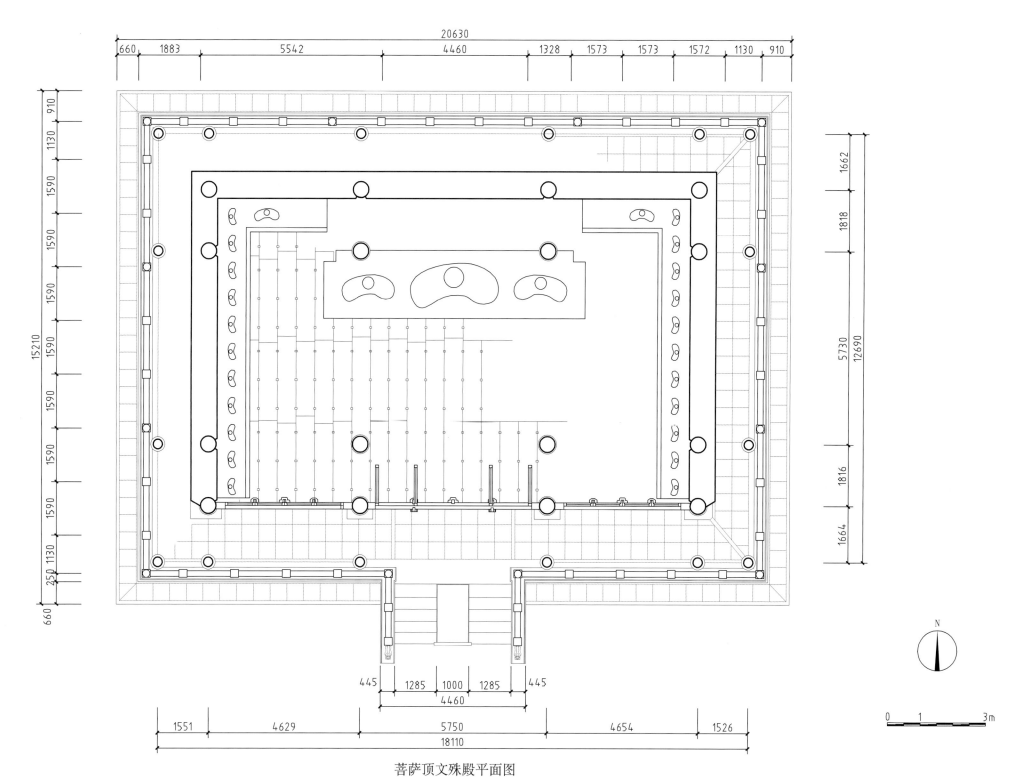

菩萨顶文殊殿平面图
Plan of Wenshudian of Bodhisattva Summit

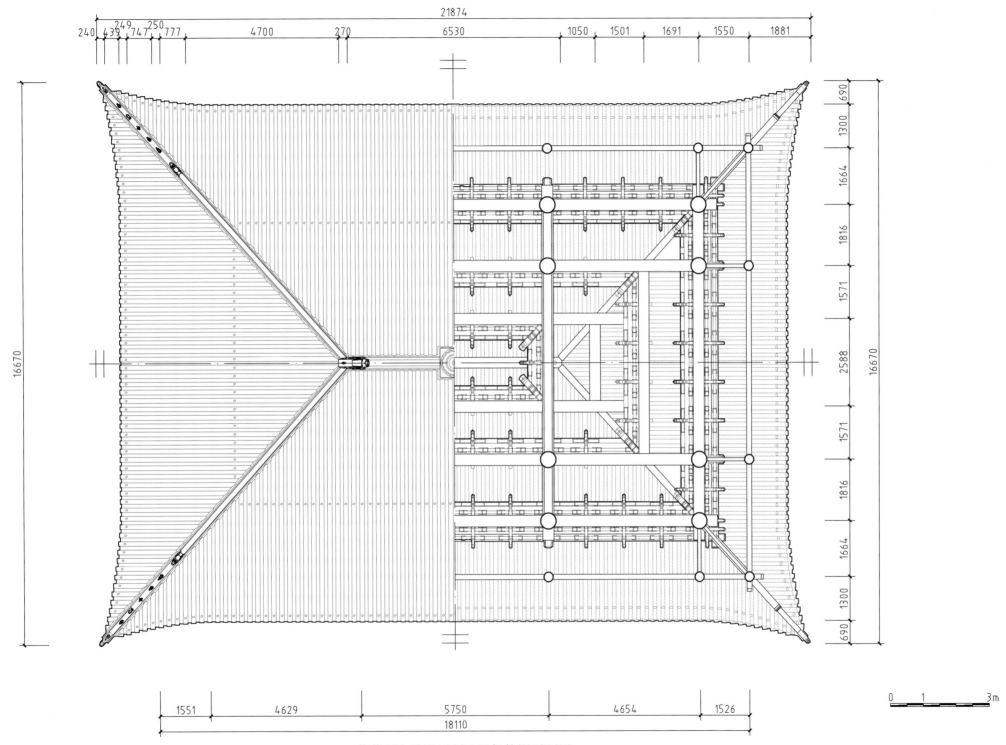

菩萨顶文殊殿屋顶及梁架仰视平面图

Plan of roof and framework of Wenshudian of Bodhisattva Summit as seen from below

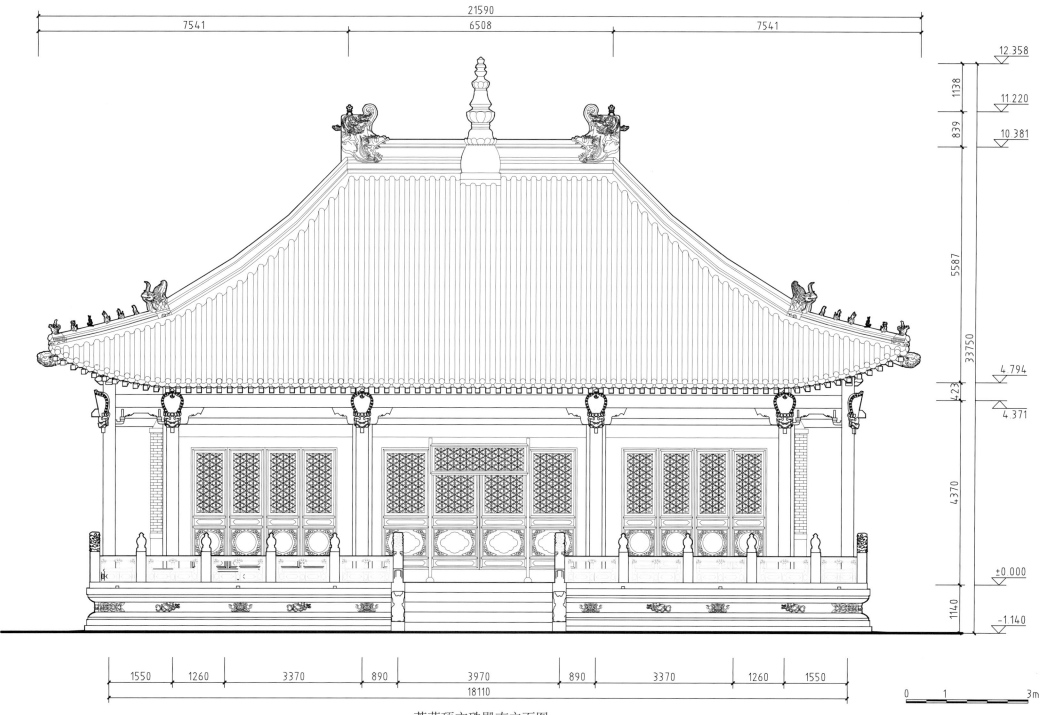

菩萨顶文殊殿南立面图
South elevation of Wenshudian of Bodhisattva Summit

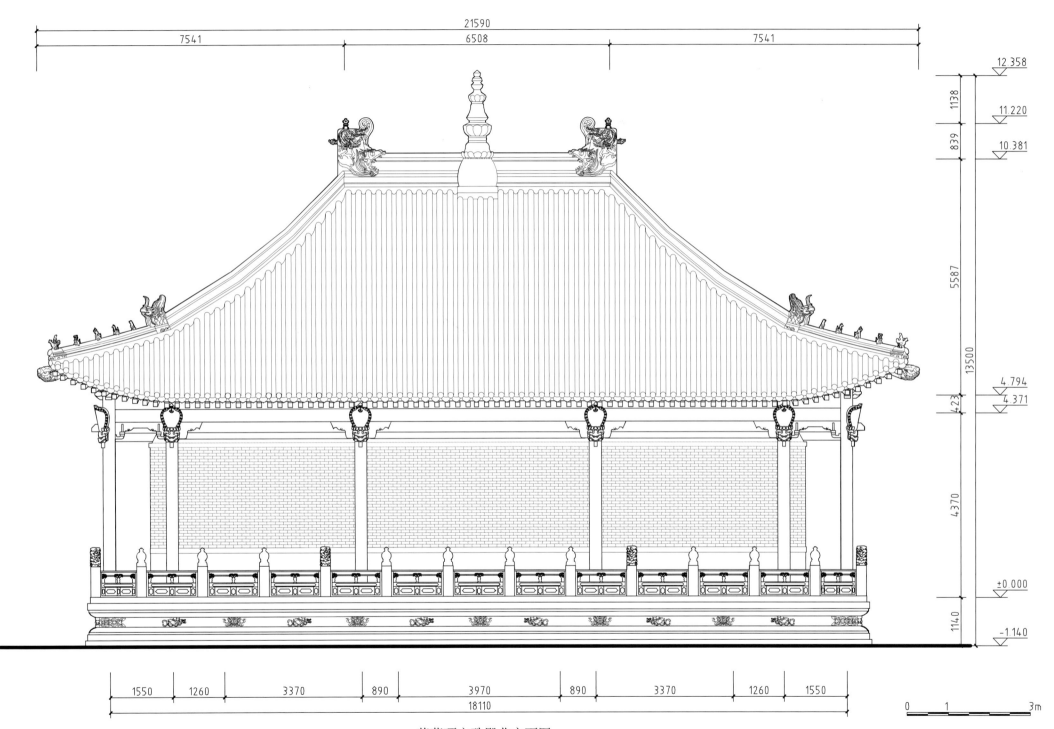

菩萨顶文殊殿北立面图

North elevation of Wenshudian of Bodhisattva Summit

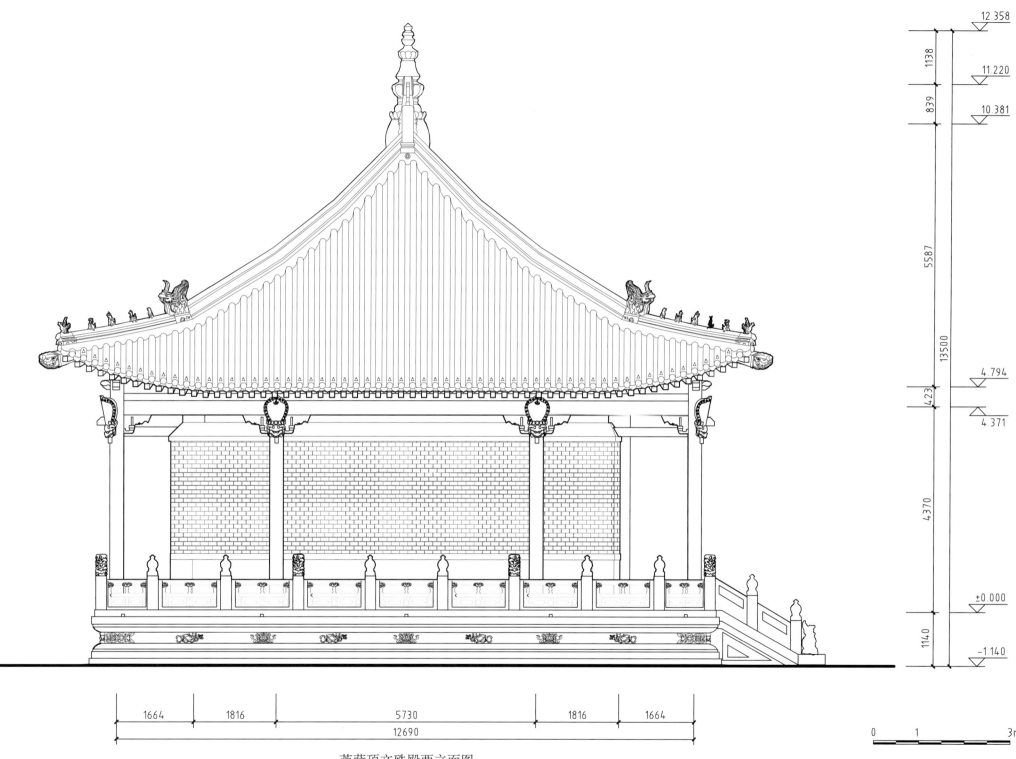

12.358

1138

11.220

839

10.381

5587

13500

4.794

423

4.371

4370

±0.000

1140

-1.140

1664　1816　5730　1816　1664

12690

0　1　3m

菩萨顶文殊殿西立面图

West elevation of Wenshudian of Bodhisattva Summit

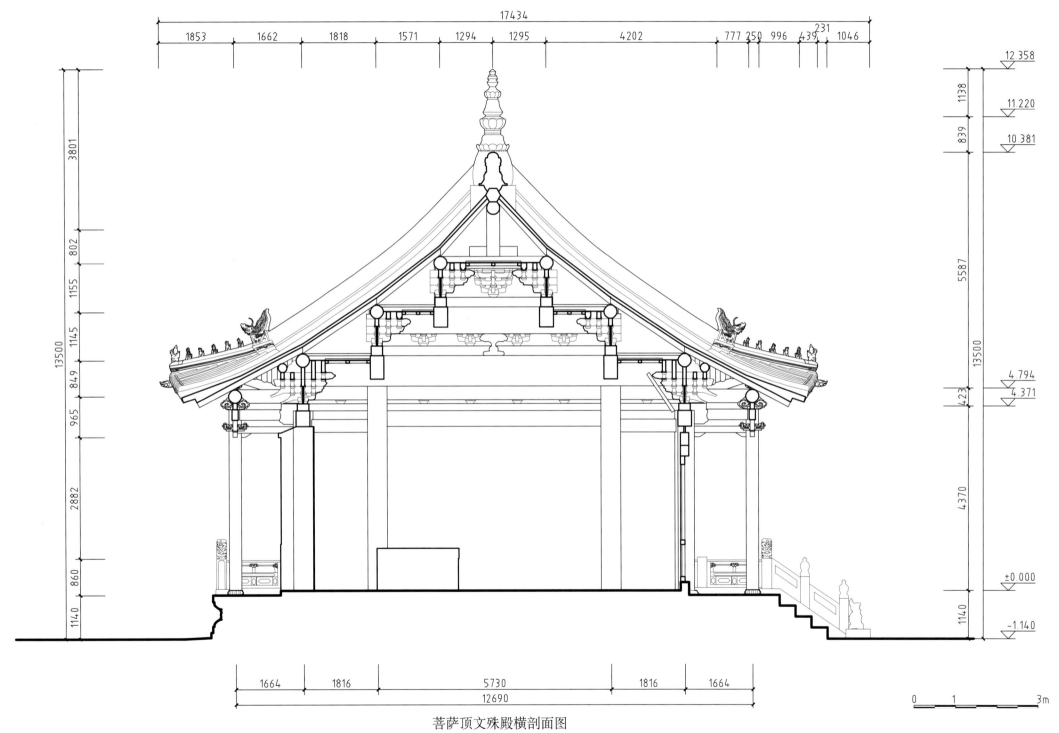

菩萨顶文殊殿横剖面图

Cross-section of Wenshudian of Bodhisattva Summit

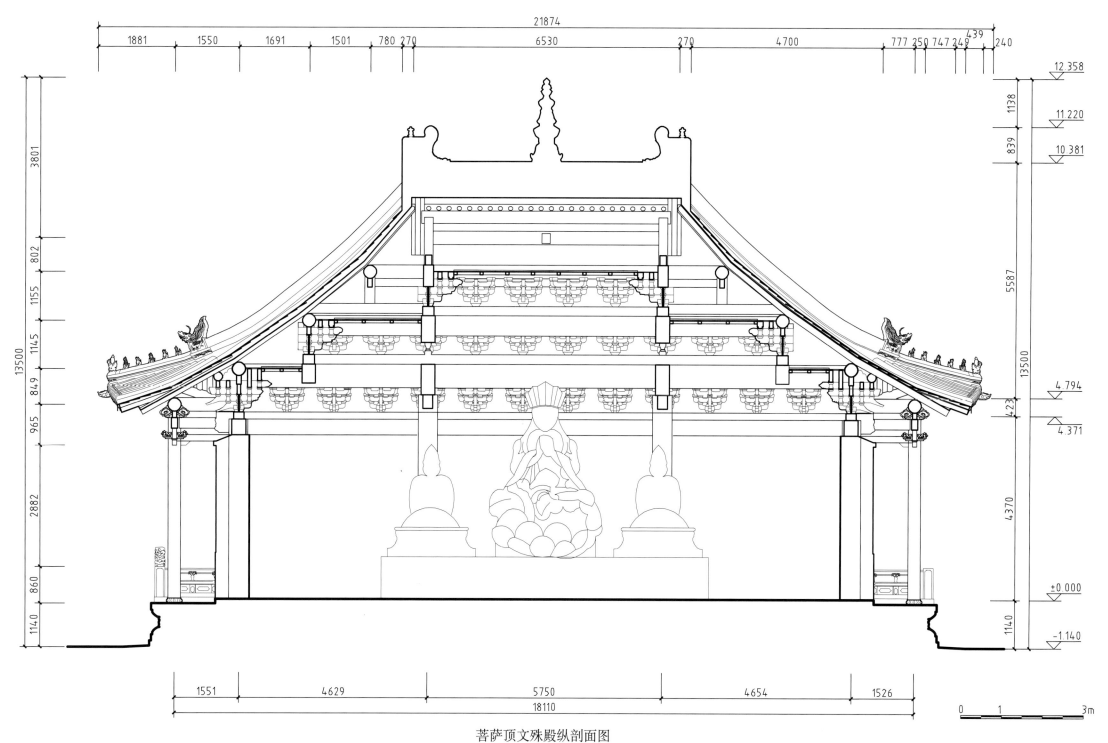

菩萨顶文殊殿纵剖面图

Longitudinal section of Wenshudian of Bodhisattva Summit

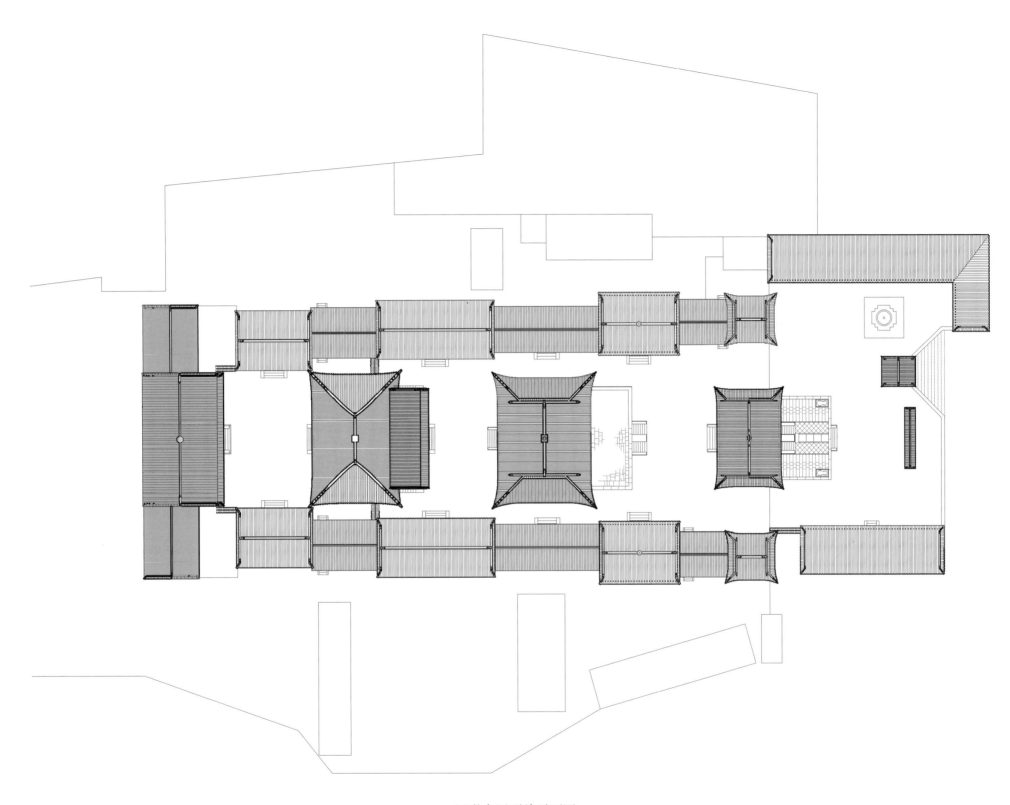

罗睺寺屋顶总平面图
Site plan of roof of Luohousi

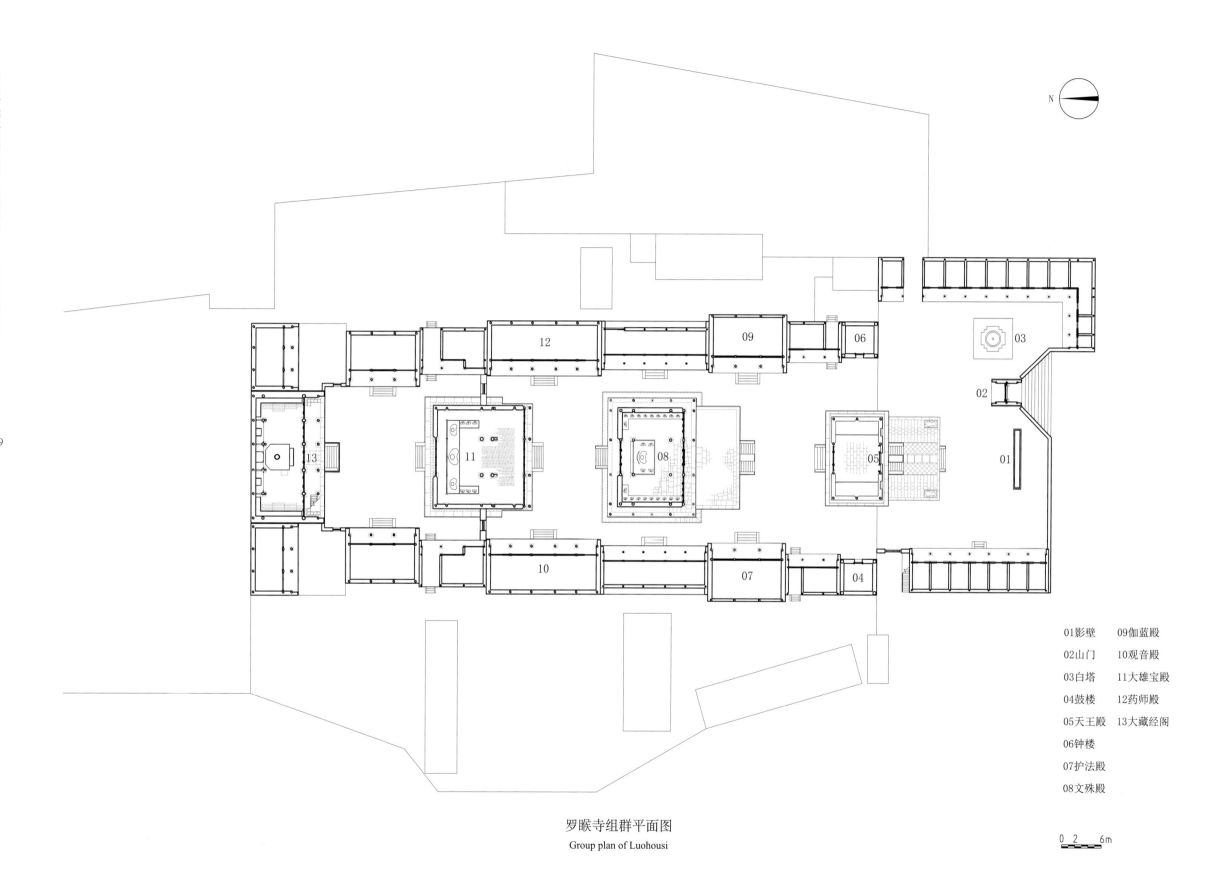

01影壁　09伽蓝殿
02山门　10观音殿
03白塔　11大雄宝殿
04鼓楼　12药师殿
05天王殿　13大藏经阁
06钟楼
07护法殿
08文殊殿

罗睺寺组群平面图
Group plan of Luohousi

0 2   6m

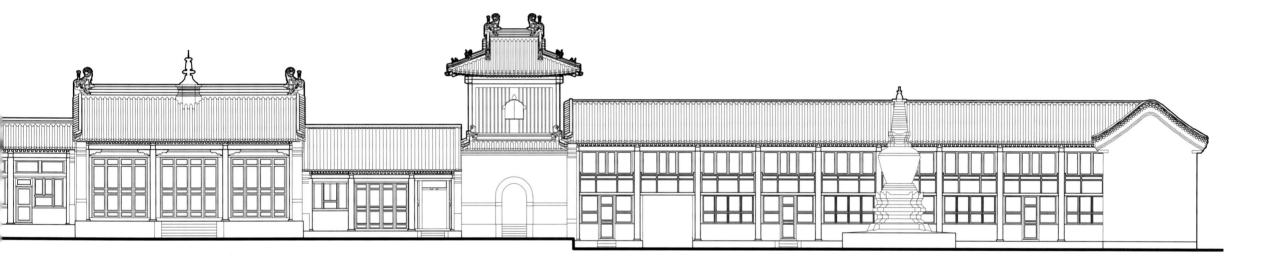

0 1   3m

罗睺寺总立面图

Site elevation of Luohousi

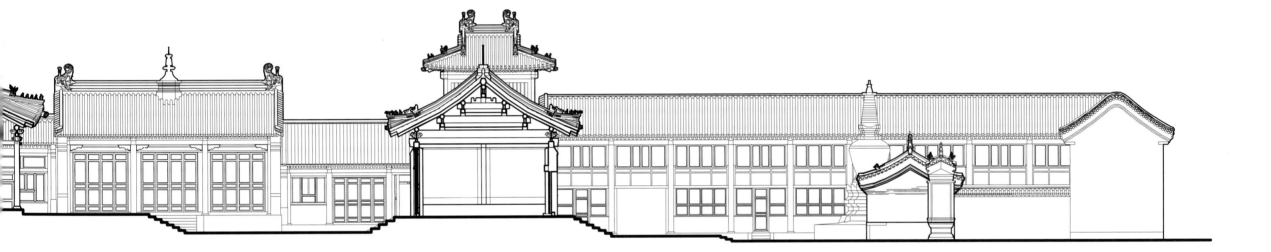

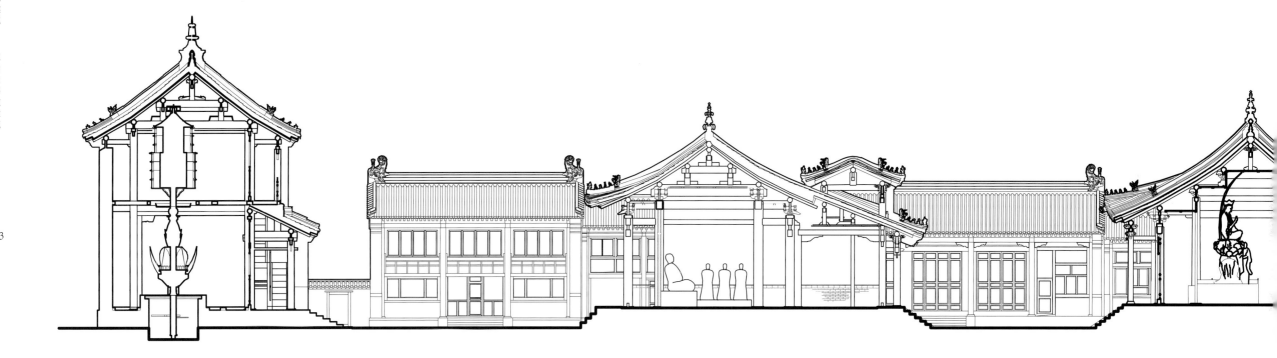

罗睺寺总剖面图

Site section of Luohousi

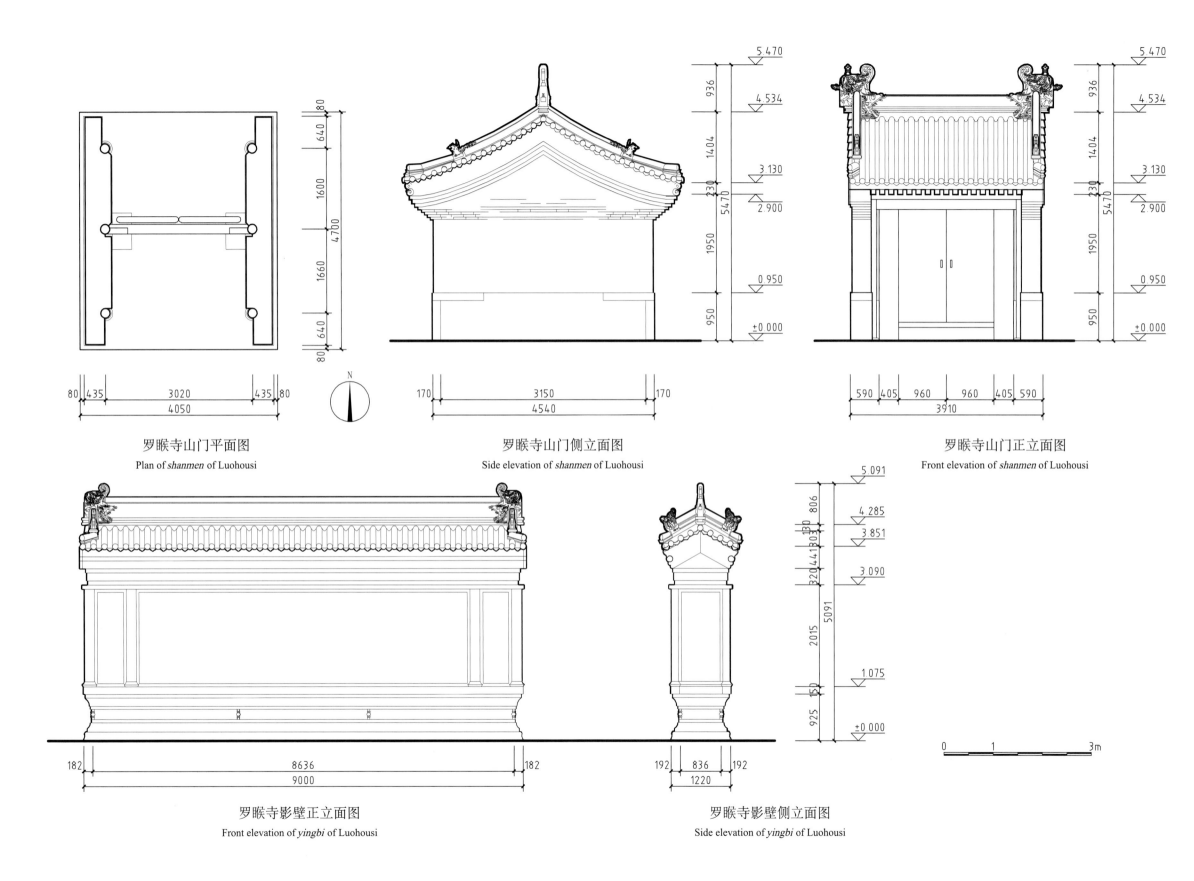

罗睺寺山门平面图
Plan of *shanmen* of Luohousi

罗睺寺山门侧立面图
Side elevation of *shanmen* of Luohousi

罗睺寺山门正立面图
Front elevation of *shanmen* of Luohousi

罗睺寺影壁正立面图
Front elevation of *yingbi* of Luohousi

罗睺寺影壁侧立面图
Side elevation of *yingbi* of Luohousi

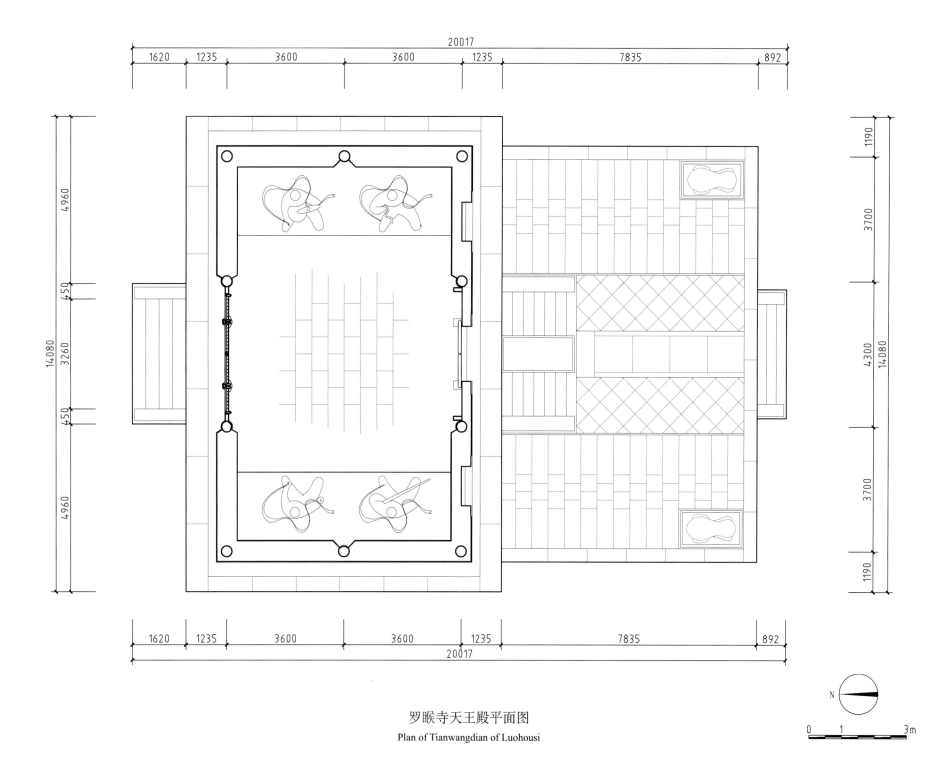

罗睺寺天王殿平面图
Plan of Tianwangdian of Luohousi

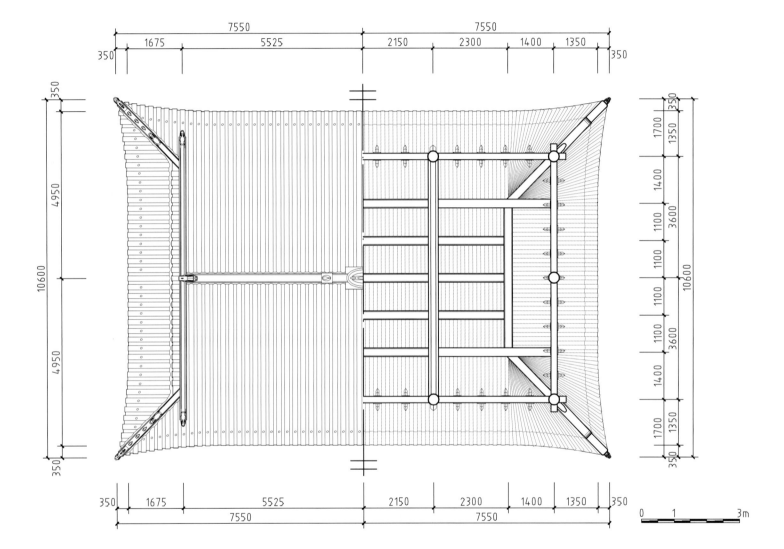

罗睺寺天王殿屋顶及梁架仰视平面图

Plan of roof and framework of Tianwangdian of Luohousi as seen from below

0   1   3m

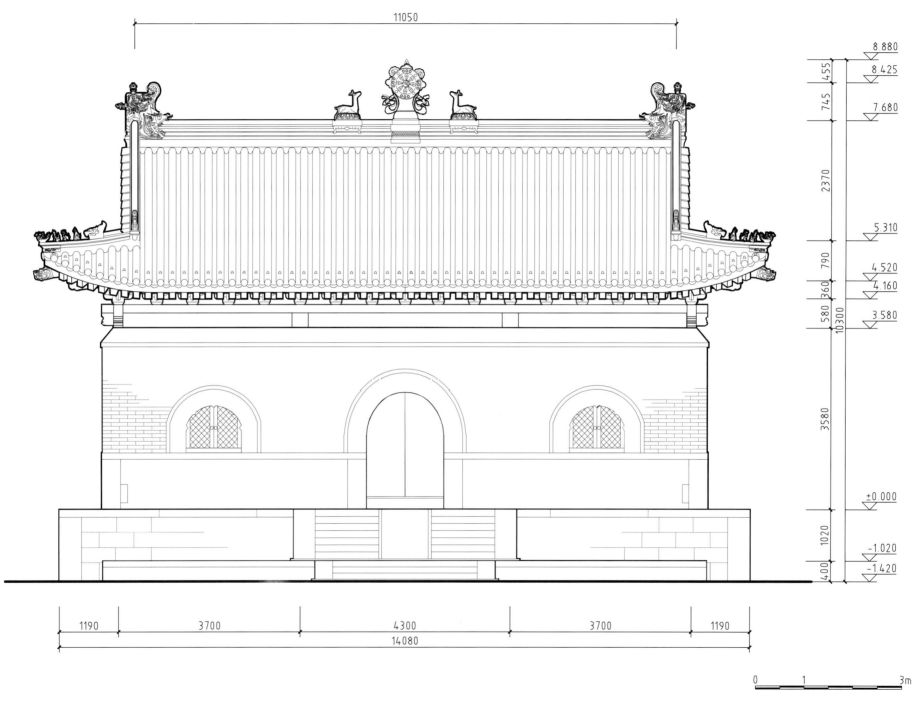

11050

8.880
8.425
455
7.680
745
2370
5.310
790
4.520
360
4.160
580
3.580
10300

3580

±0.000

1020
-1.020

400
-1.420

1190   3700   4300   3700   1190
14080

0   1   3m

罗睺寺天王殿正立面图
Front elevation of Tianwangdian of Luohousi

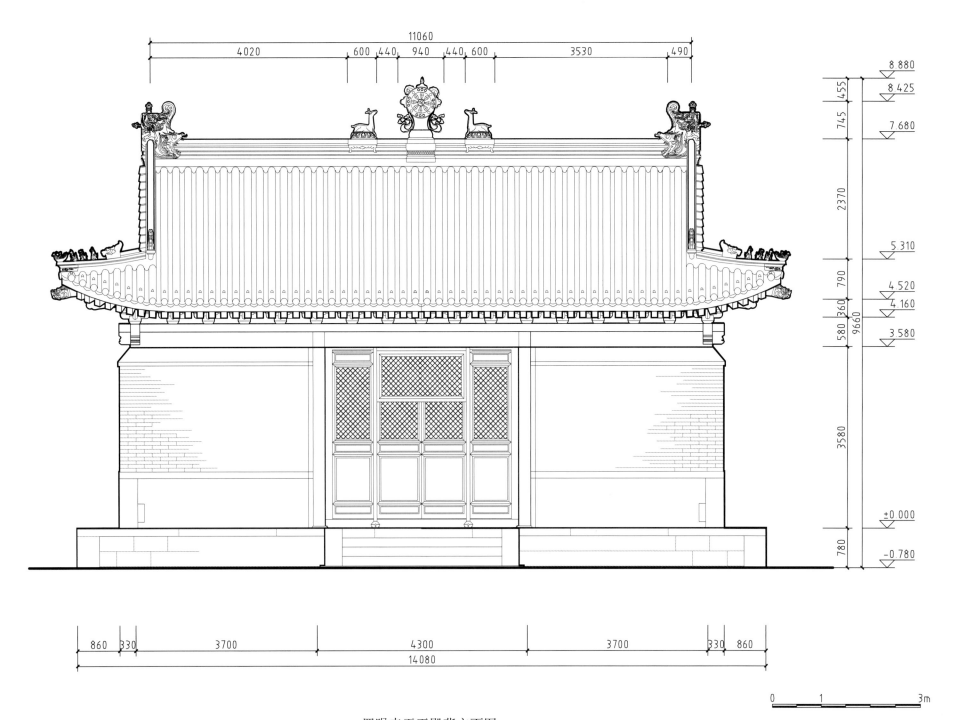

罗睺寺天王殿背立面图

Rear elevation of Tianwangdian of Luohousi

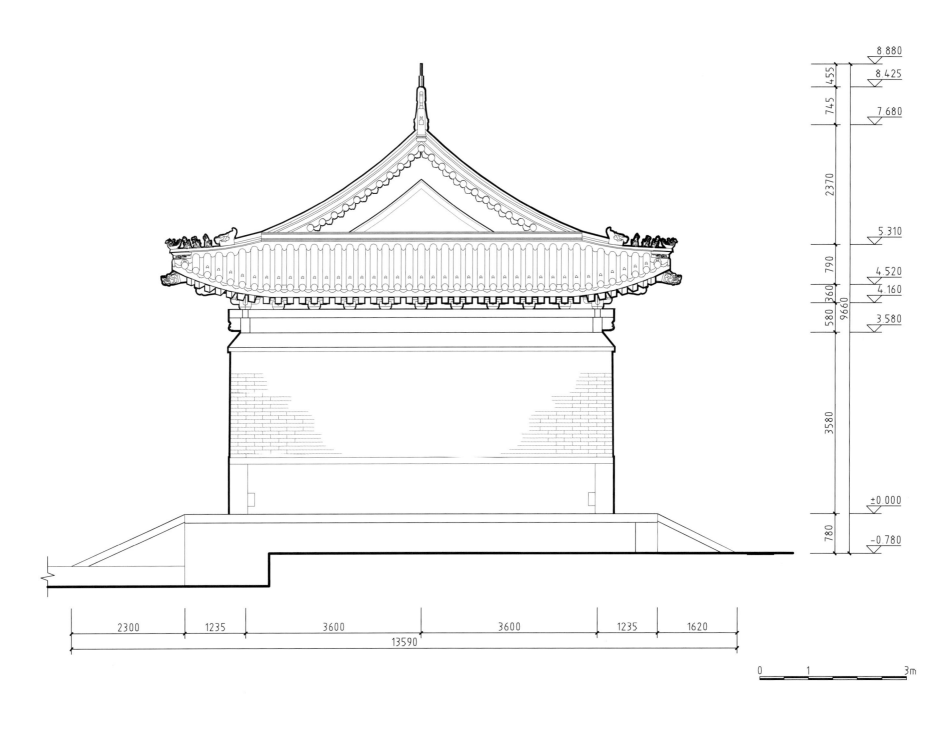

8 880
4 55
8 425
745
7 680
2370
5 310
790
4 520
360
4 160
580
9 660
3 580
3 580
±0.000
780
-0.780

2300 | 1235 | 3600 | 3600 | 1235 | 1620
13590

0　　　1　　　3m

罗睺寺天王殿侧立面图

Side elevation of Tianwangdian of Luohousi

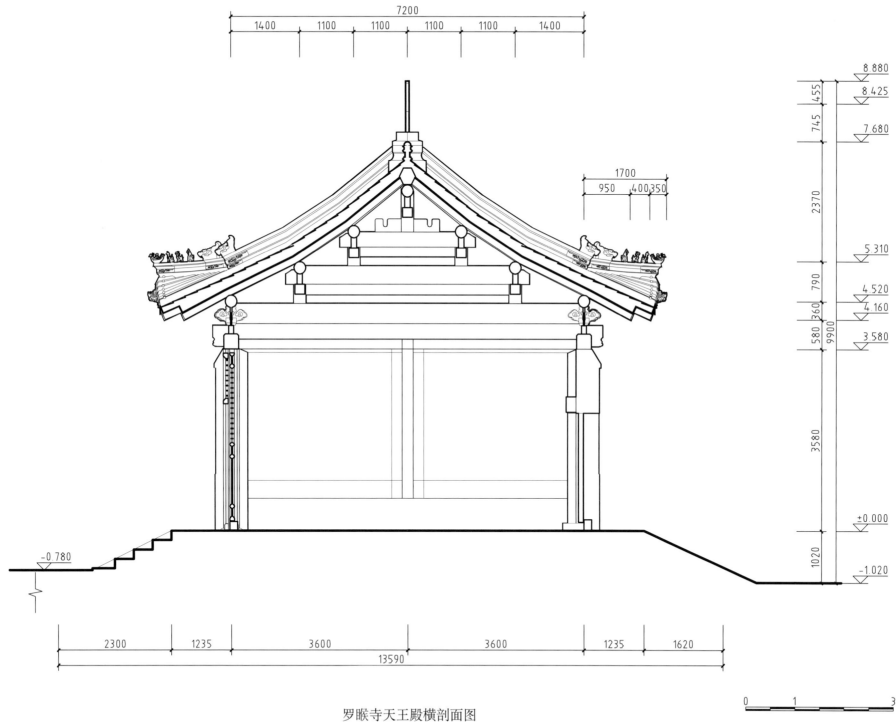

罗睺寺天王殿横剖面图

Cross-section of Tianwangdian of Luohousi

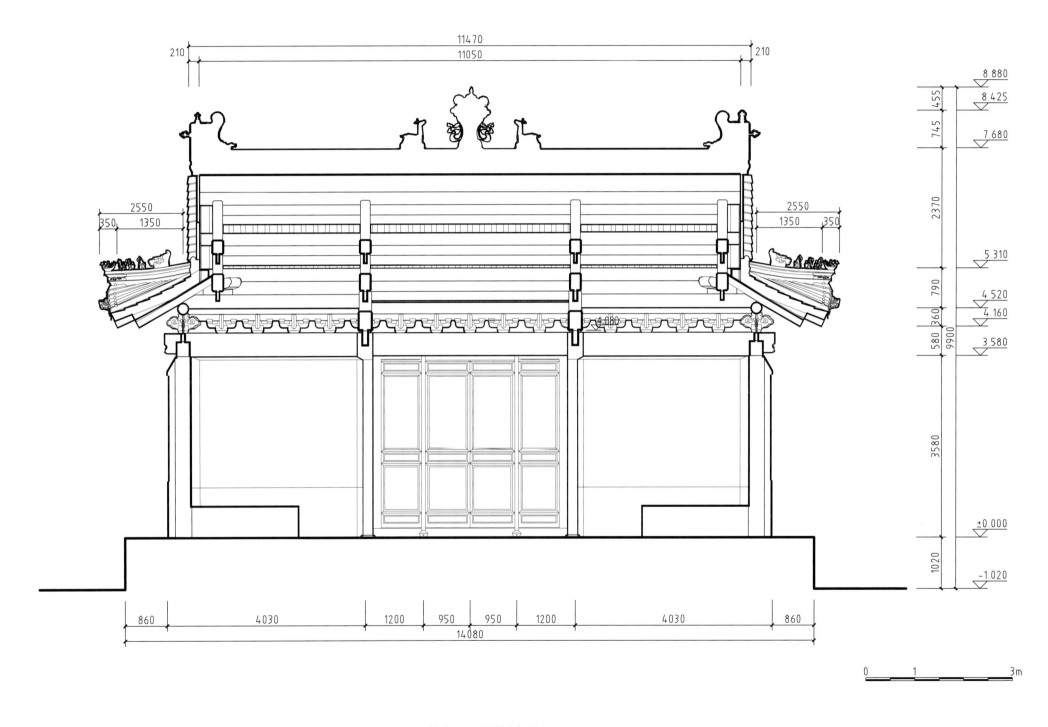

罗睺寺天王殿纵剖面图

Longitudinal section of Tianwangdian of Luohousi

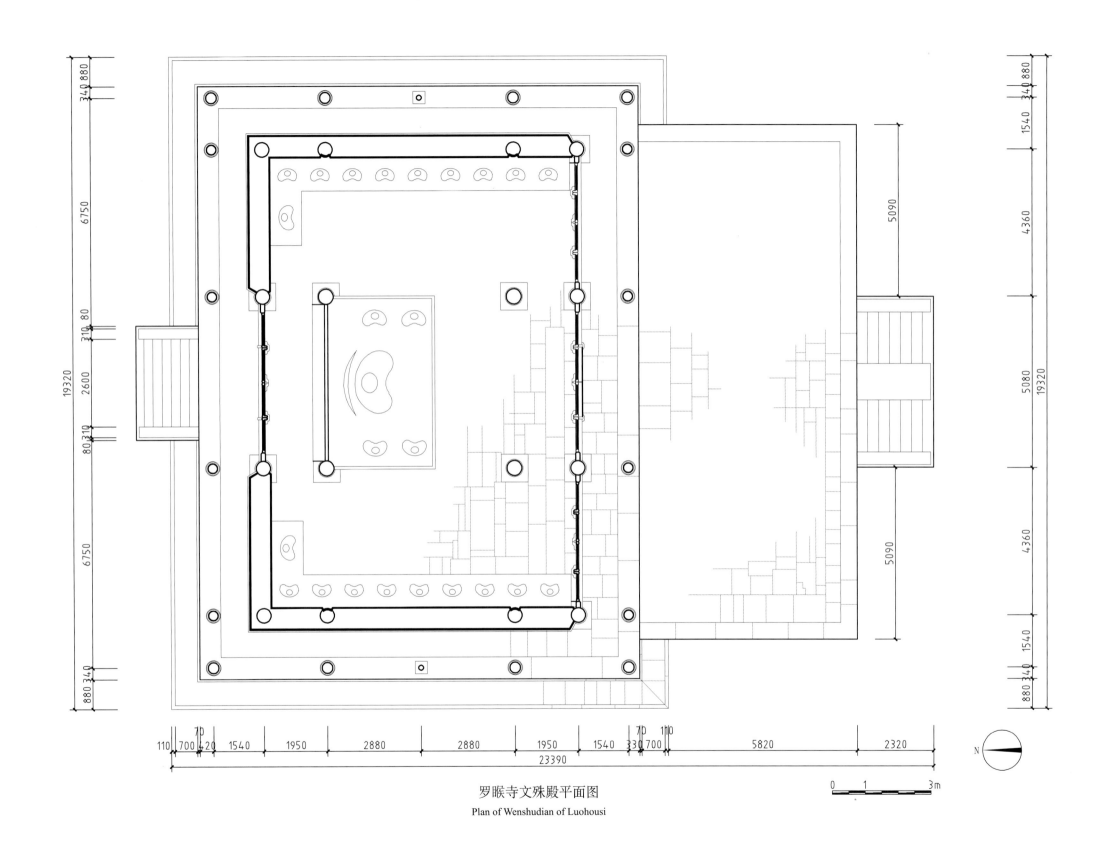

罗睺寺文殊殿平面图

Plan of Wenshudian of Luohousi

0 1 3m

N

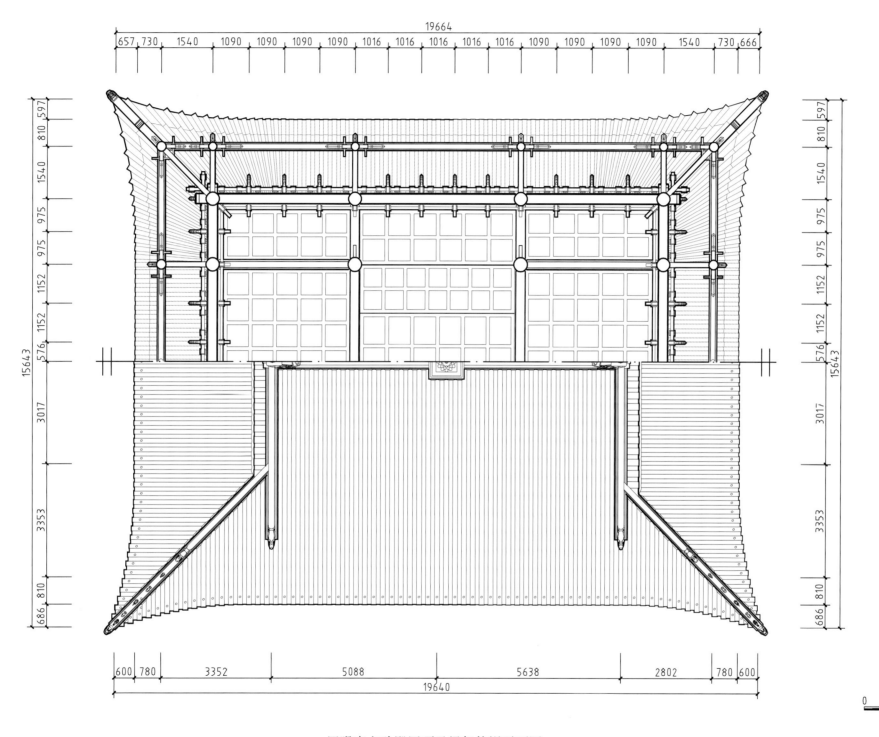

19664

657 730 1540 1090 1090 1090 1090 1016 1016 1016 1016 1016 1090 1090 1090 1090 1540 730 666

597 810 1540 975 975 1152 1152 576 3017 3353 810 686

15643

597 810 1540 975 975 1152 1152 576 3017 3353 810 686

15643

600 780 3352 5088 5638 2802 780 600

19640

0 1 3m

罗睺寺文殊殿屋顶及梁架仰视平面图

Plan of roof and framework of Wenshudian of Luohousi as seen from below

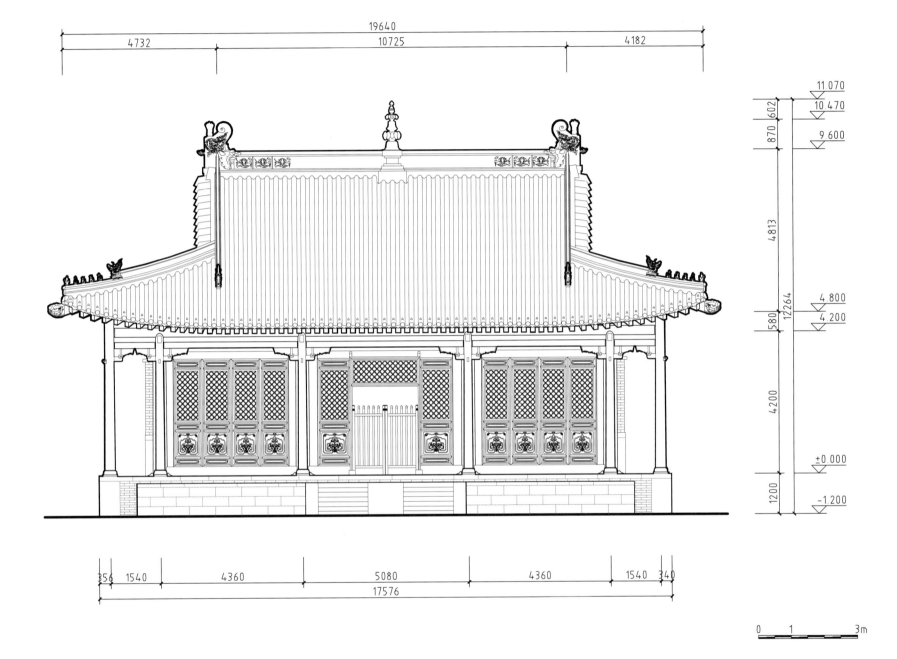

19640

4732　　10725　　4182

11.070
10.470
9.600

602
870
4813
4.800
580
12264
4.200
4200
±0.000
1200
-1.200

356 1540　4360　5080　4360　1540 340
17576

0　1　3m

罗睺寺文殊殿正立面图

Front elevation of Wenshudian of Luohousi

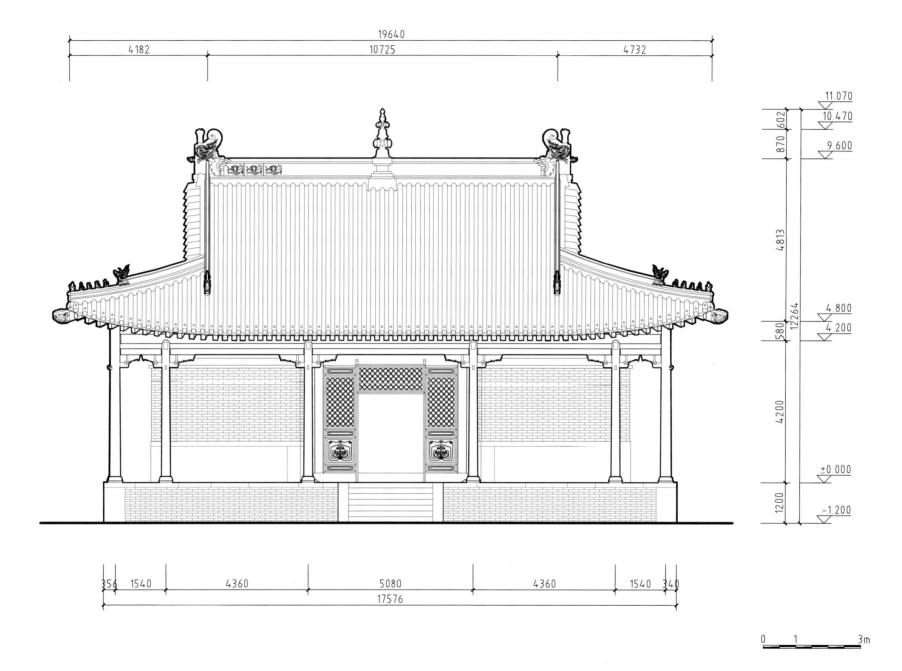

罗睺寺文殊殿背立面图
Rear elevation of Wenshudian of Luohousi

0   1   3m

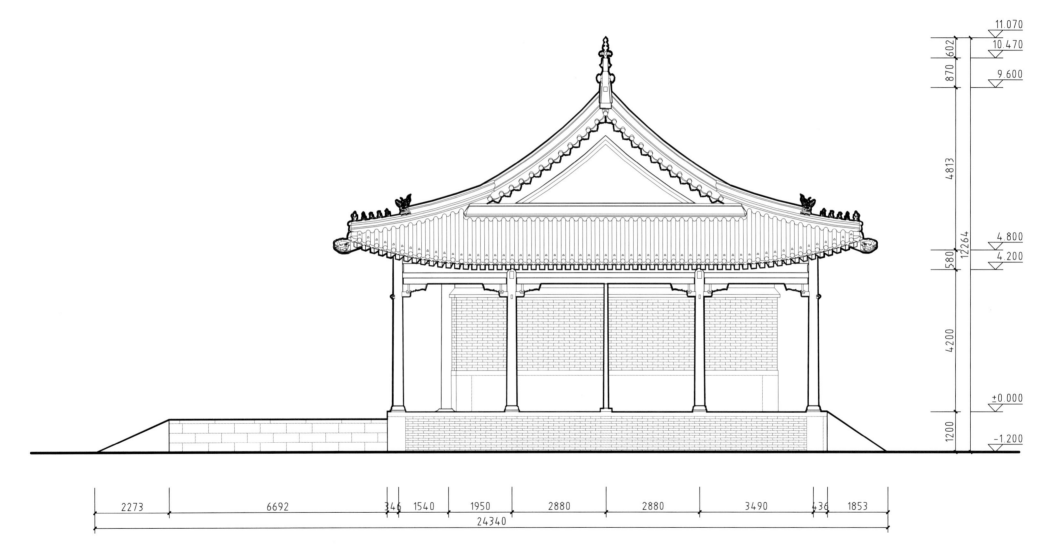

11.070

10.470

9.600

4.800

4.200

±0.000

-1.200

602

870

4813

580

12264

4200

1200

2273　6692　346　1540　1950　2880　2880　3490　436　1853

24340

0　1　3m

罗睺寺文殊殿侧立面图

Side elevation of Wenshudian of Luohousi

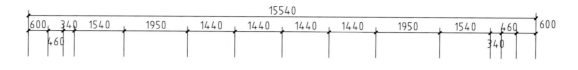

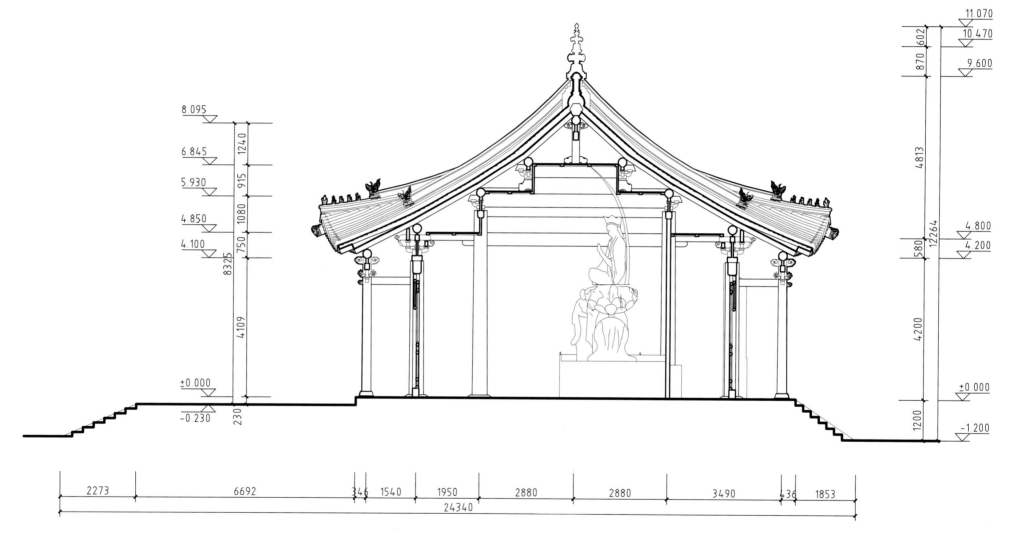

罗睺寺文殊殿横剖面图
Cross-section of Wenshudian of Luohousi

0 1 3m

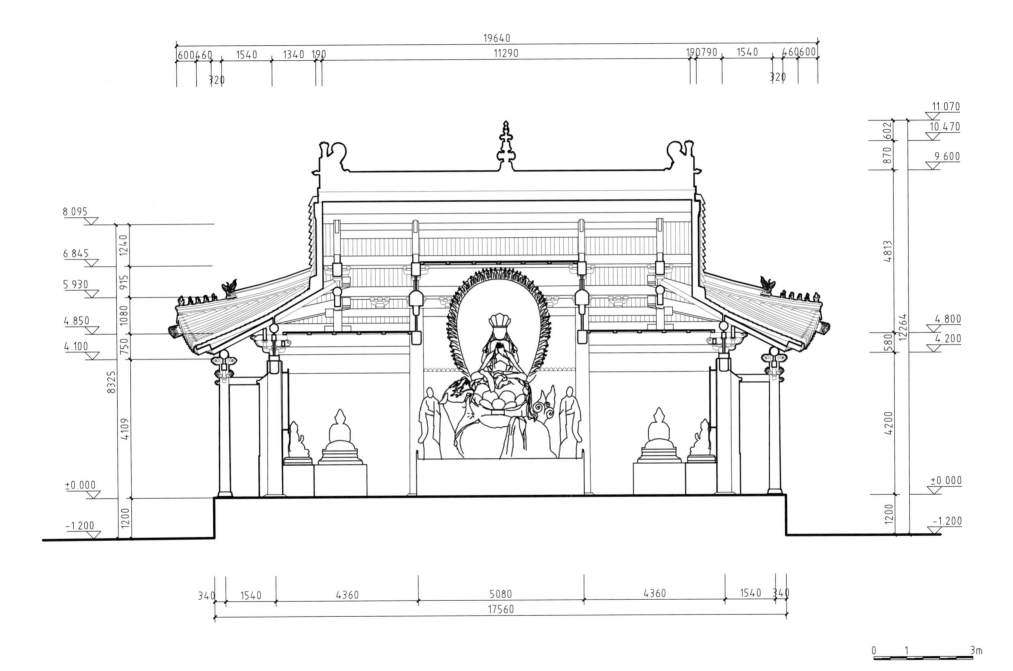

罗睺寺文殊殿纵剖面图

Longitudinal section of Wenshudian of Luohousi

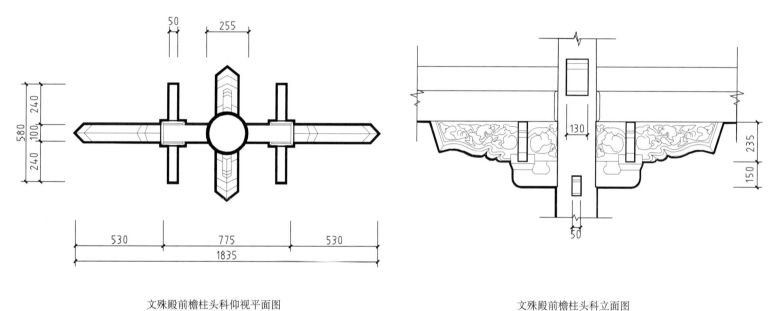

文殊殿前檐柱头科仰视平面图

文殊殿前檐柱头科立面图

0　　　　　　0.5　　　　　　1m

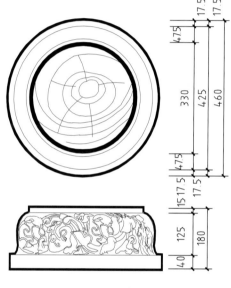

文殊殿前檐柱础大样

0　　　　0.25　　　　0.5m

罗睺寺文殊殿大样图（一）

Detail of Wenshudian of Luohousi (I)

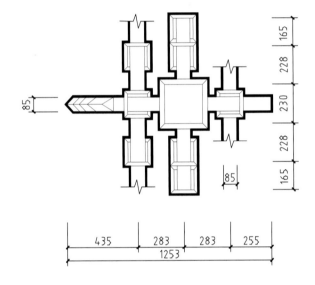

文殊殿下金平身科仰视平面图

文殊殿下金平身科侧立面图

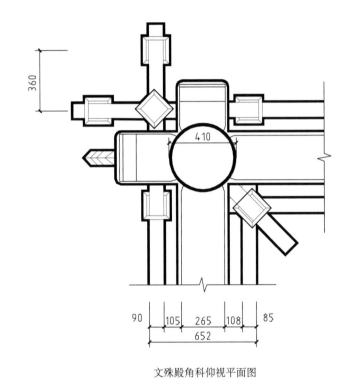

文殊殿角科仰视平面图

罗睺寺文殊殿大样图（二）

Detail of Wenshudian of Luohousi (II)

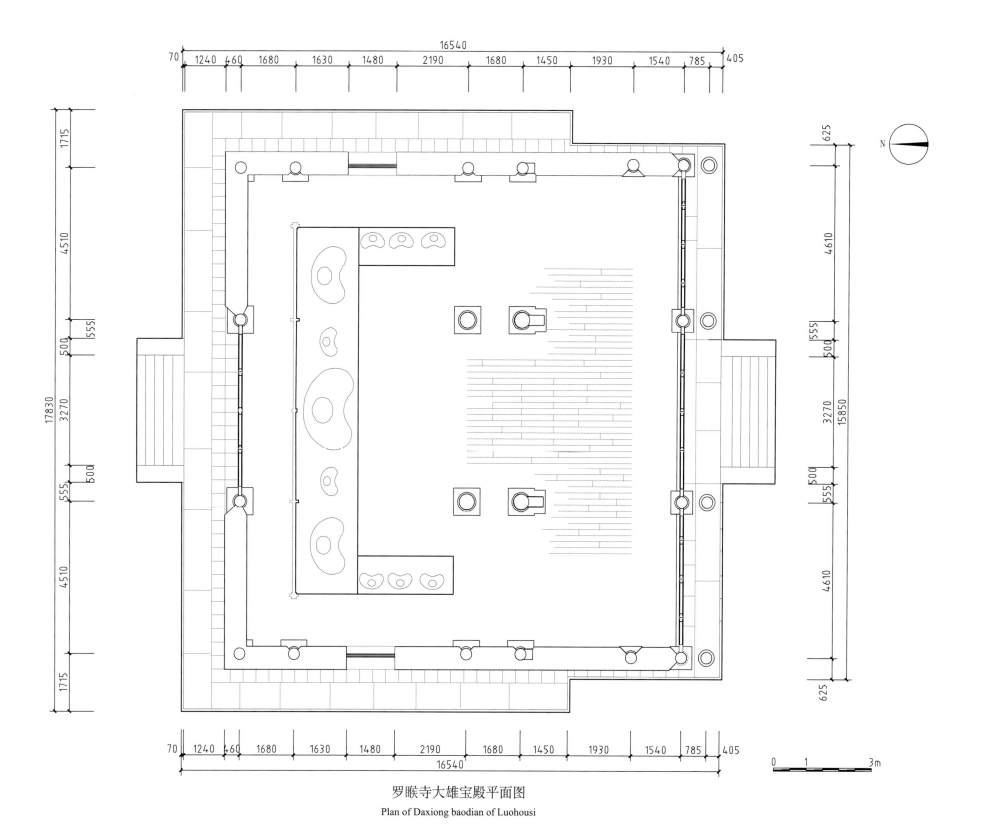

罗睺寺大雄宝殿平面图
Plan of Daxiong baodian of Luohousi

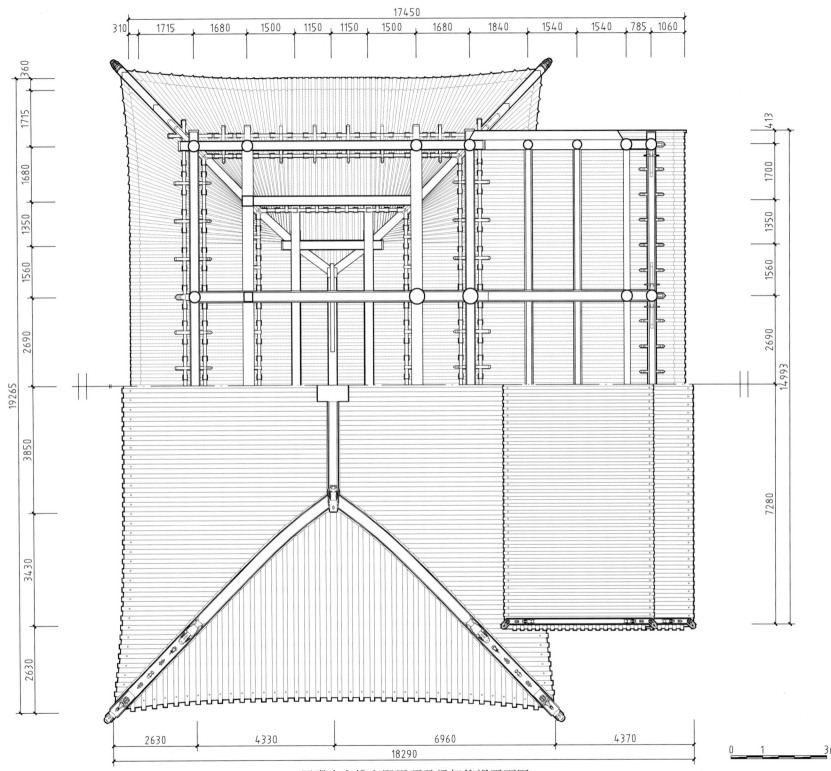

罗睺寺大雄宝殿屋顶及梁架仰视平面图

Plan of roof and framework of Daxiong baodian of Luohousi as seen from below

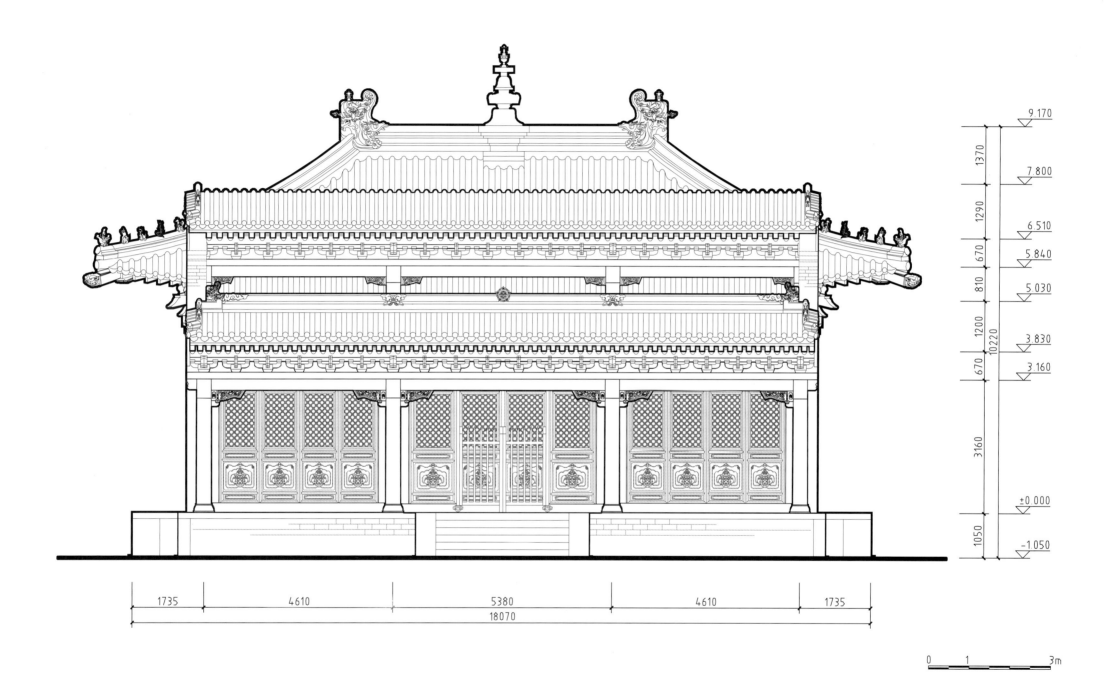

罗睺寺大雄宝殿正立面图

Front elevation of Daxiong baodian of Luohousi

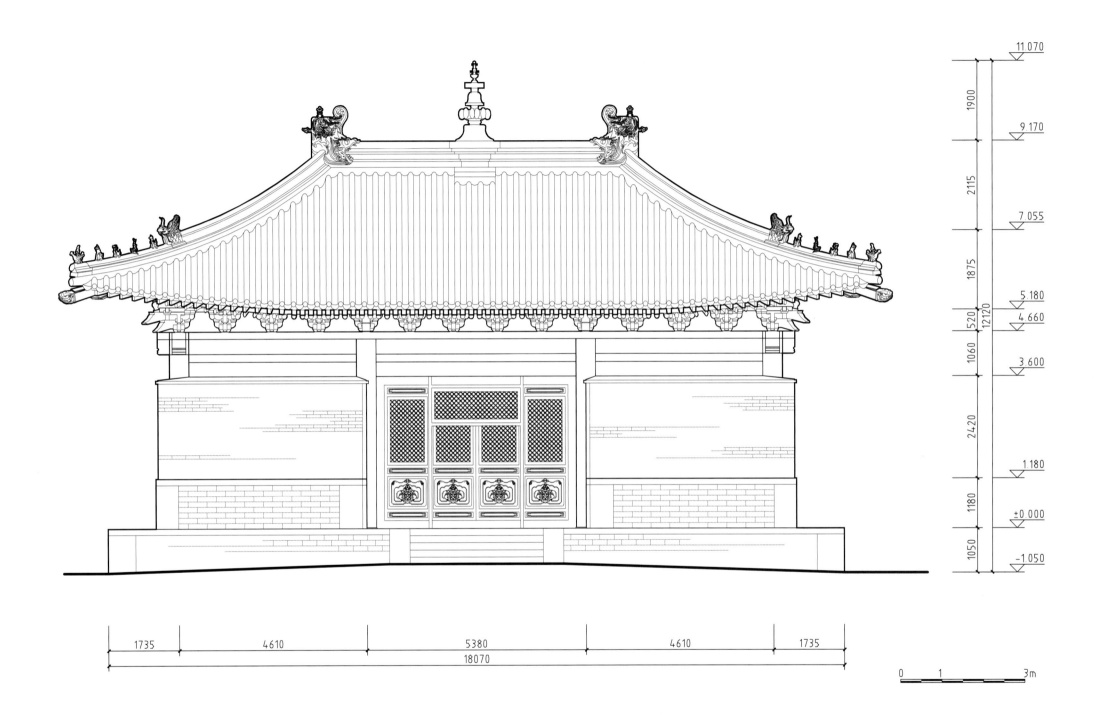

11.070

1900

9.170

2115

7.055

1875

5.180

520 12120

4.660

1060

3.600

2420

1.180

1180

±0.000

1050

-1.050

1735 4610 5380 4610 1735
18070

0 1 3m

罗睺寺大雄宝殿背立面图
Rear elevation of Daxiong baodian of Luohousi

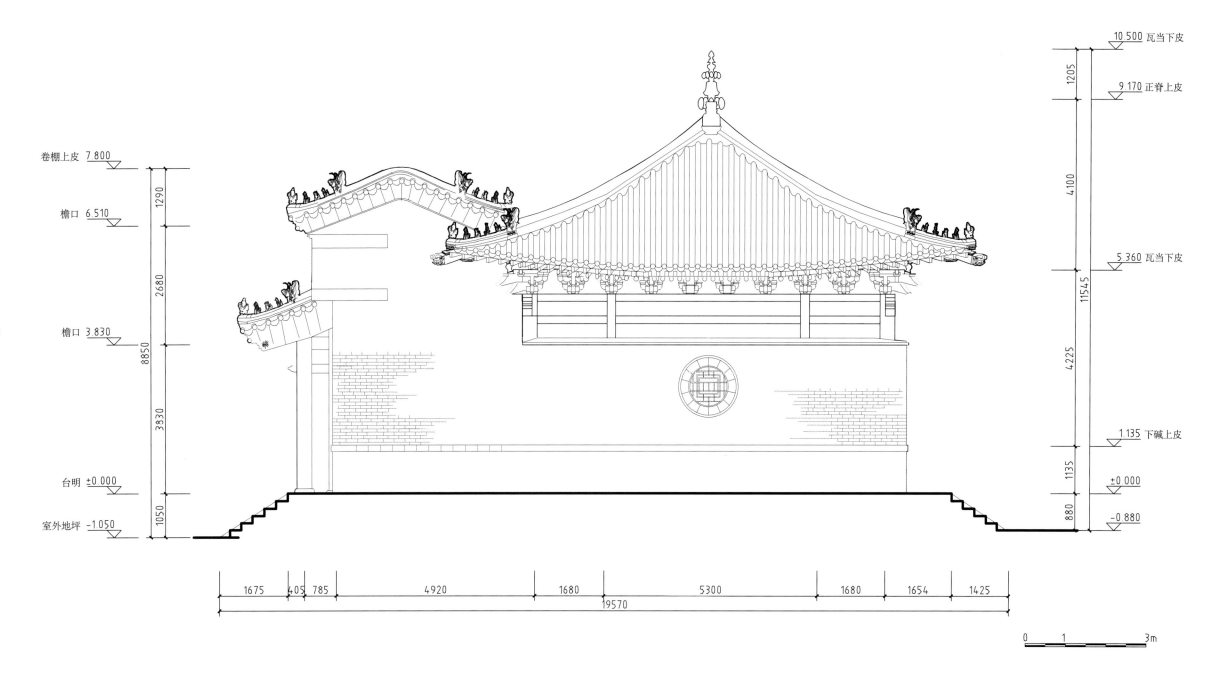

卷棚上皮 7.800

檐口 6.510

檐口 3.830

台明 ±0.000

室外地坪 -1.050

1290

2680

3830

8850

1050

10.500 瓦当下皮

9.170 正脊上皮

1205

4100

5.360 瓦当下皮

11545

4225

1.135 下碱上皮

1135

±0.000

880

-0.880

1675　405　785　　4920　　1680　　5300　　1680　1654　1425

19570

0　1　　　3m

罗睺寺大雄宝殿侧立面图

Side elevation of Daxiong baodian of Luohousi

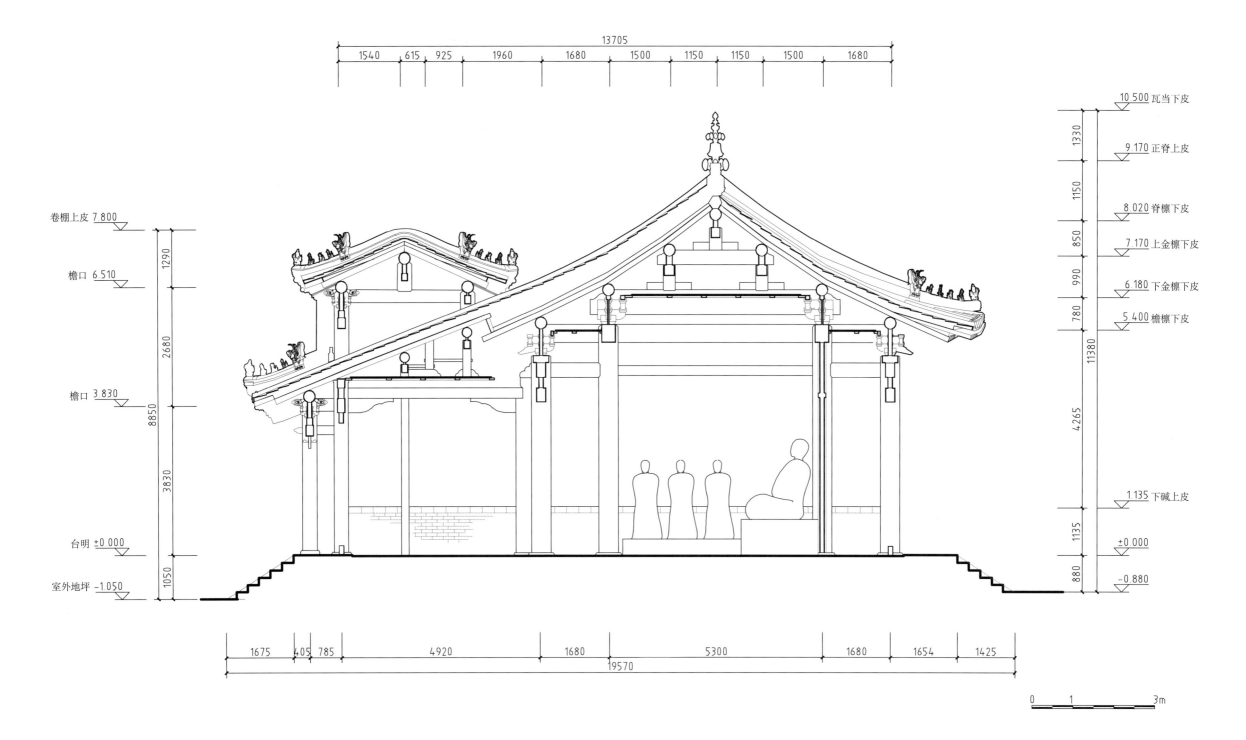

13705

1540 615 925 1960 1680 1500 1150 1150 1500 1680

10 500 瓦当下皮
1330
9 170 正脊上皮
1150
8 020 脊檩下皮
850
7 170 上金檩下皮
990
6 180 下金檩下皮
780
5 400 檐檩下皮
11380
4 265
1 135 下碱上皮
1135
±0.000
880
-0.880

卷棚上皮 7.800
1290
檐口 6.510
2680
檐口 3.830
8850
3830
台明 ±0.000
1050
室外地坪 -1.050

1675 405 785 4920 1680 5300 1680 1654 1425
19570

0 1 3m

罗睺寺大雄宝殿横剖面图
Cross-section of Daxiong baodian of Luohousi

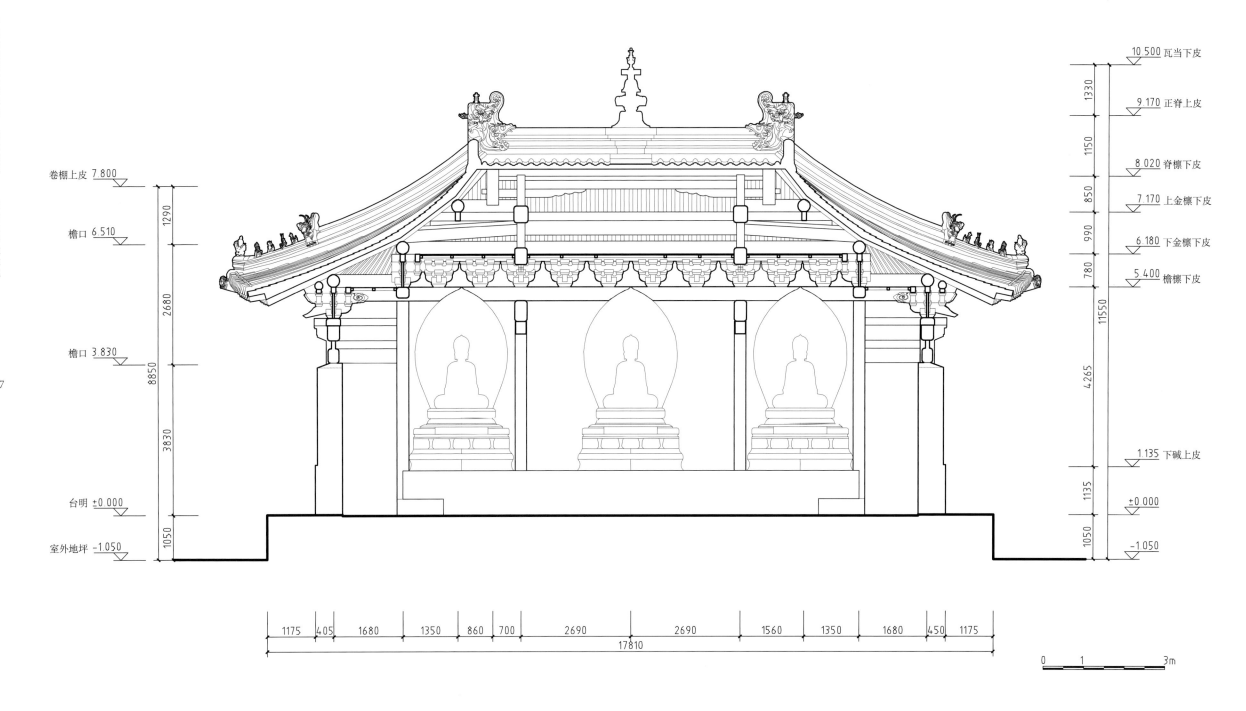

卷棚上皮 7.800

檐口 6.510

檐口 3.830

台明 ±0.000

室外地坪 -1.050

10.500 瓦当下皮

9.170 正脊上皮

8.020 脊檩下皮

7.170 上金檩下皮

6.180 下金檩下皮

5.400 檐檩下皮

1.135 下碱上皮

±0.000

-1.050

1290

2680

3830

1050

8850

1330

1150

850

990

780

11550

4.265

1135

1050

1175 | 405 | 1680 | 1350 | 860 | 700 | 2690 | 2690 | 1560 | 1350 | 1680 | 450 | 1175

17810

0 1 3m

罗睺寺大雄宝殿纵剖面图

Longitudinal section of Daxiong baodian of Luohousi

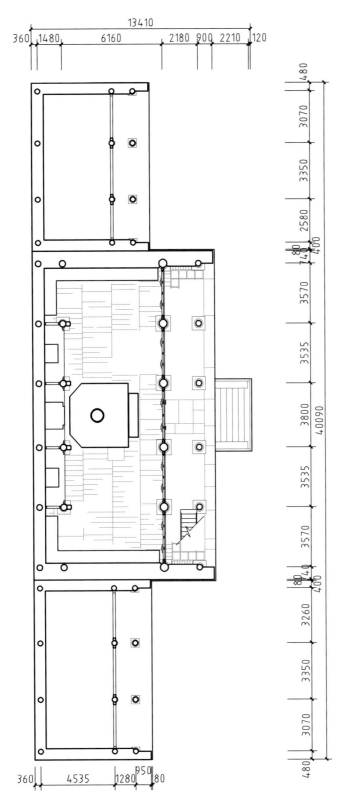

罗睺寺大藏经阁一层平面图
Plan of first floor of *cangjingge* of Luohousi

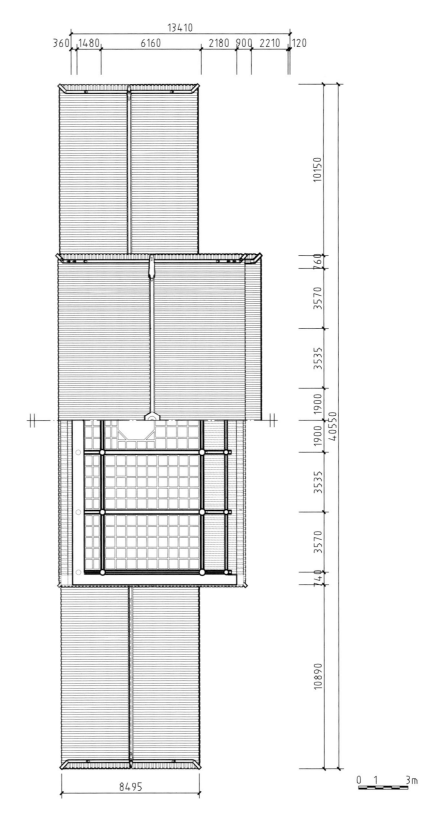

罗睺寺大藏经阁屋顶平面图及二层梁架仰视平面图
Plan of roof and second floor framework of *cangjingge* of Luohousi as seen from below

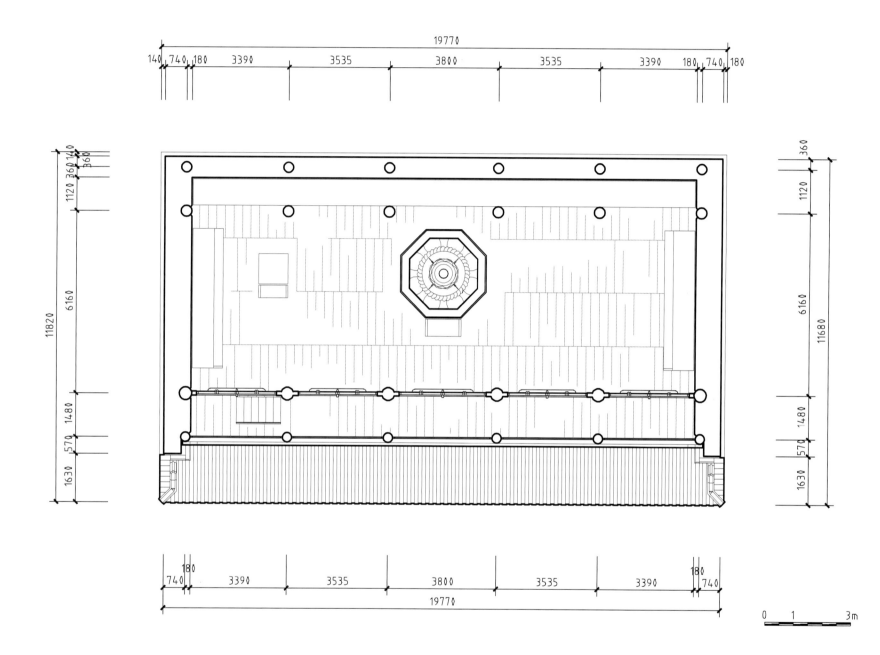

罗睺寺大藏经阁二层平面图
Plan of second floor *cangjingge* of Luohousi

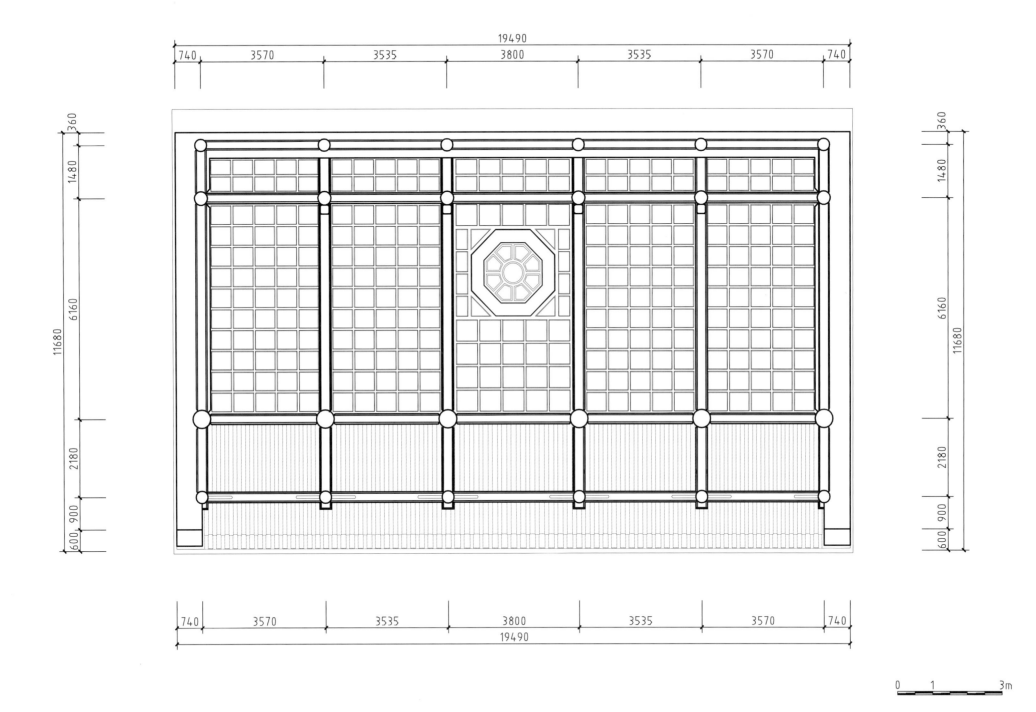

19490

740 3570 3535 3800 3535 3570 740

360
1480
6160
11680
2180
600 900

360
1480
6160
11680
2180
600 900

740 3570 3535 3800 3535 3570 740
19490

0 1 3m

罗睺寺大藏经阁一层梁架仰视平面图

Plan of first floor framework of *cangjingge* of Luohousi as seen from below

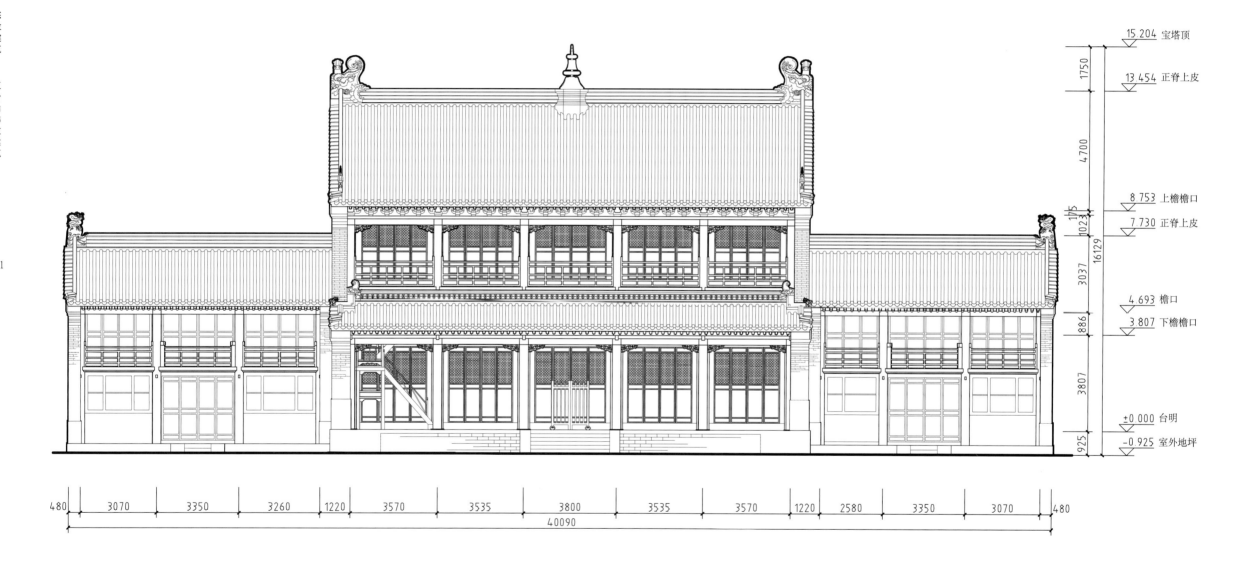

15 204 宝塔顶

13 454 正脊上皮

8 753 上檐檐口

7 730 正脊上皮

4 693 檐口

3 807 下檐檐口

+0.000 台明

-0.925 室外地坪

1750

4 700

175

1023

3037

16129

886

3807

925

上檐檐口

下檐檐口

480 3070 3350 3260 1220 3570 3535 3800 3535 3570 1220 2580 3350 3070 480

40090

0 1 3m

罗睺寺大藏经阁南立面图

South elevation of *cangjingge* of Luohousi

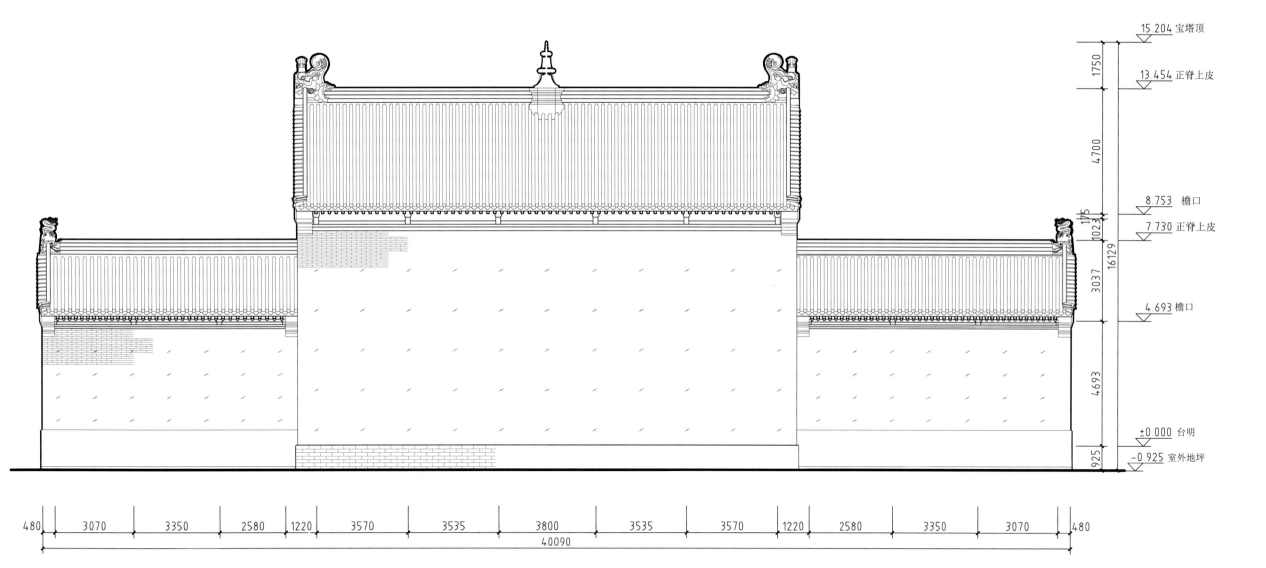

15.204 宝塔顶

1750

13.454 正脊上皮

4700

8.753 檐口

175

7.730 正脊上皮

1023

16129

3037

4.693 檐口

4693

±0.000 台明

925

-0.925 室外地坪

480　3070　3350　2580　1220　3570　3535　3800　3535　3570　1220　2580　3350　3070　480

40090

0　1　3m

罗睺寺大藏经阁北立面图

North elevation of *cangjingge* of Luohousi

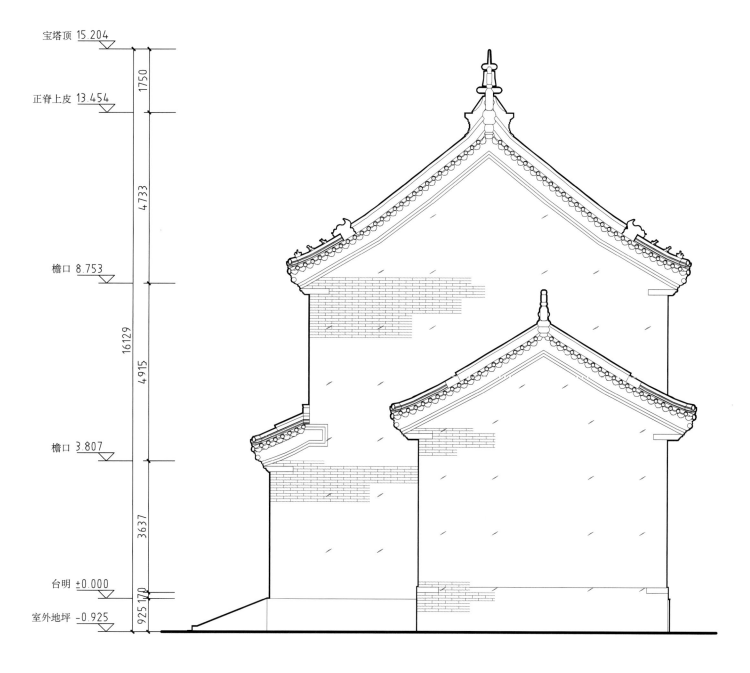

宝塔顶 15.204

正脊上皮 13.454

1750

4733

檐口 8.753

16129

4915

檐口 3.807

3637

台明 ±0.000

925 170

室外地坪 -0.925

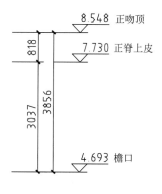

8.548 正吻顶

818

7.730 正脊上皮

3037

3856

4.693 檐口

0 1 3m

罗睺寺大藏经阁东立面图

East elevation of *cangjingge* of Luohousi

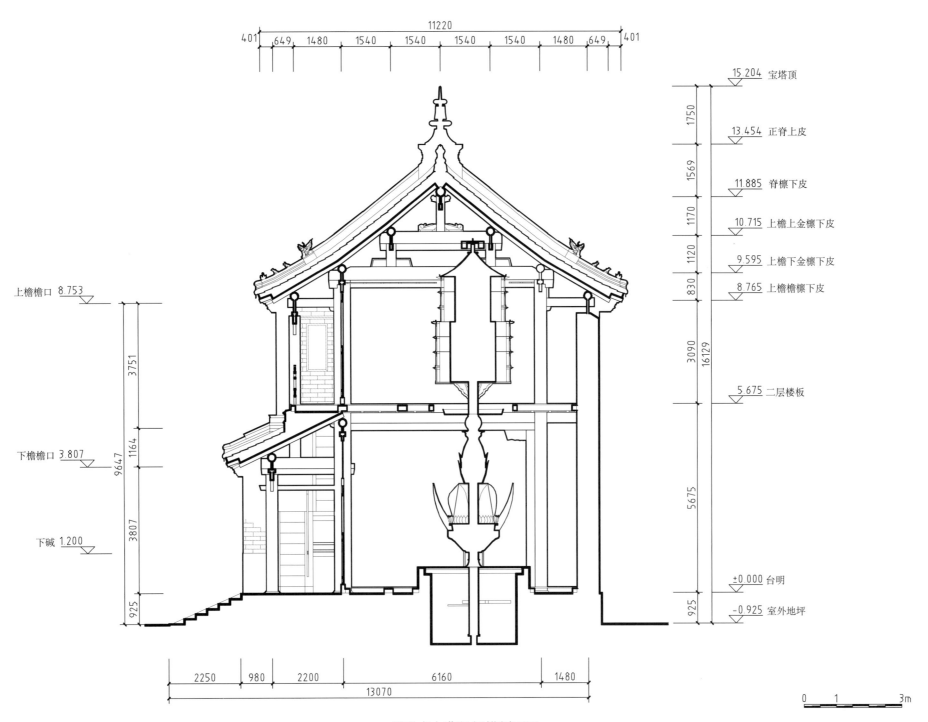

罗睺寺大藏经阁横剖面图
Cross-section of *cangjingge* of Luohousi

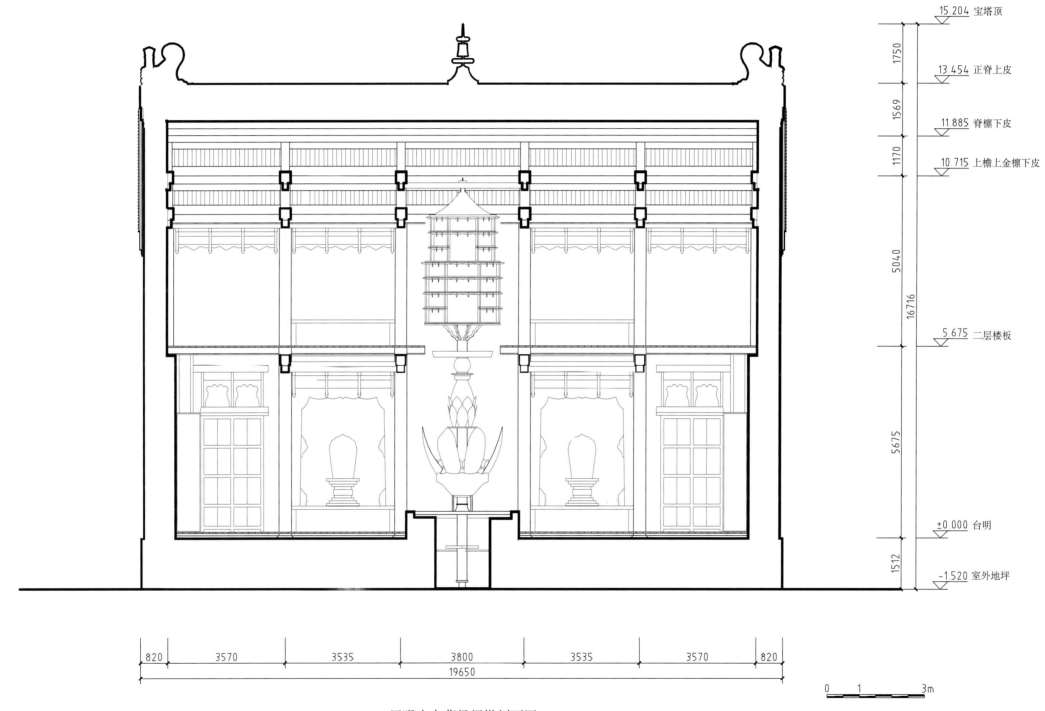

15.204 宝塔顶

1750

13.454 正脊上皮

1569

11.885 脊檩下皮

1170

10.715 上檐上金檩下皮

5040

16716

5.675 二层楼板

5675

±0.000 台明

1512

-1.520 室外地坪

820　　3570　　3535　　3800　　3535　　3570　　820

19650

0　1　3m

罗睺寺大藏经阁纵剖面图

Longitudinal section of *cangjingge* of Luohousi

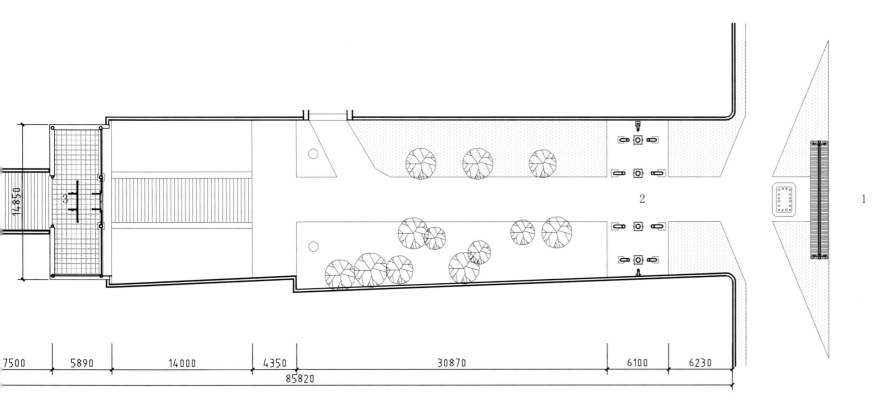

14850

7500　5890　14000　4350　30870　6100　6230

85820

1　影壁
2　牌楼
3　前殿
4　天王殿
5　延寿殿
6　大白塔
7　大藏经阁

N

0 1　　5m

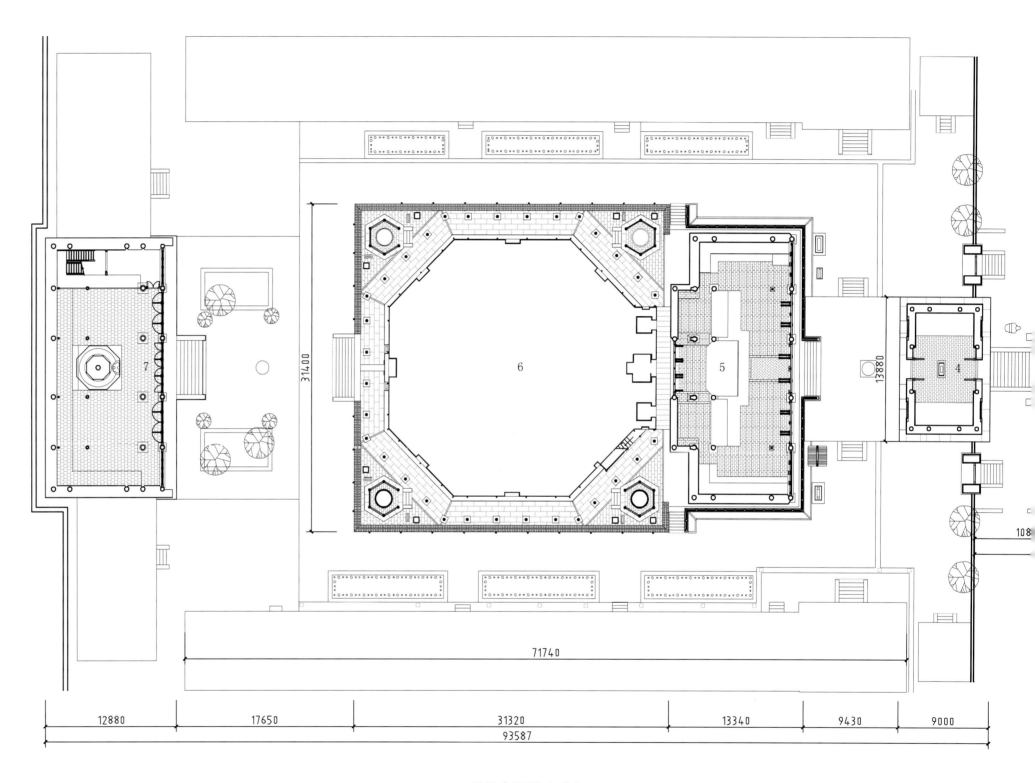

31400

71740

13880

12880　17650　31320　13340　9430　9000

93587

塔院寺组群平面图
Group plan of Tayuansi

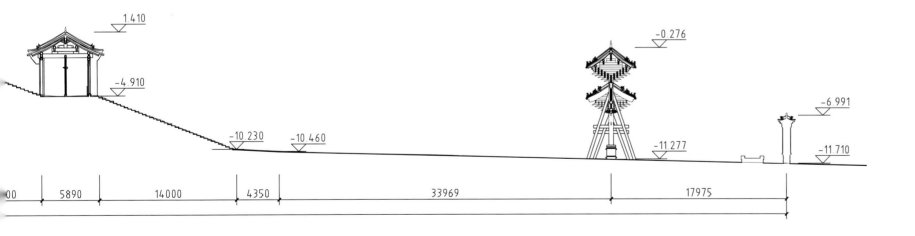

1.410

-0.276

-4.910

-6.991

-10.230   -10.460

-11.277

-11.710

5890   14000   4350   33969   17975

00

0 1   5m

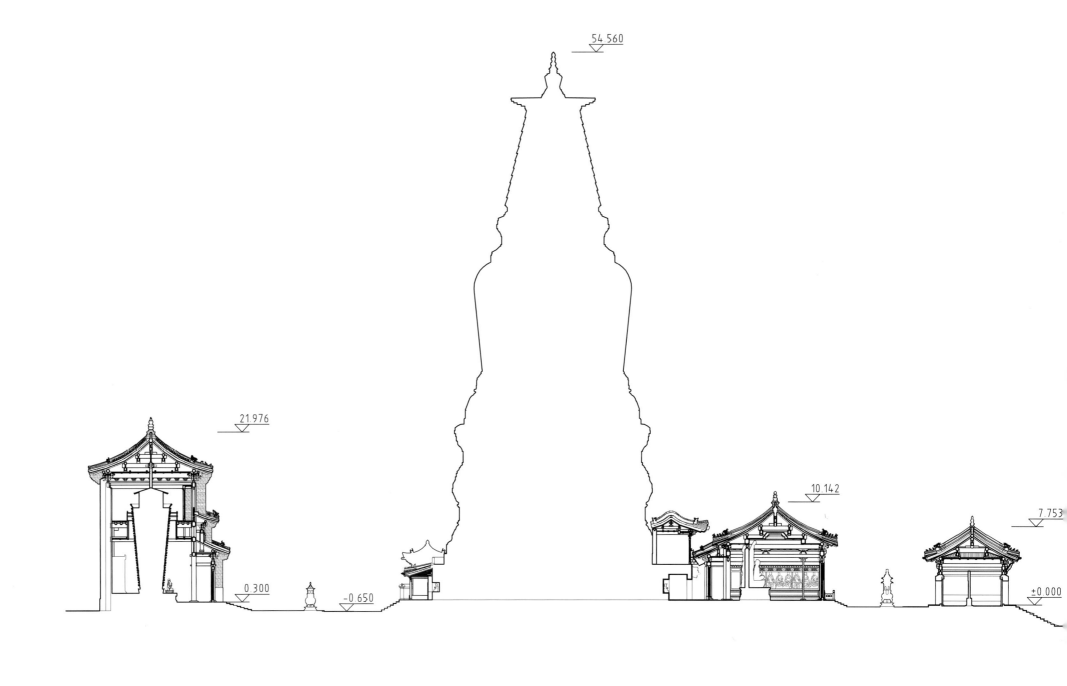

54.560

21.976

10.142

7.753

0.300

-0.650

±0.000

| 12347 | 17650 | 31320 | 13340 | 9430 | 9000 | 94 |

186225

塔院寺总剖面图

Site section of Tayuansi

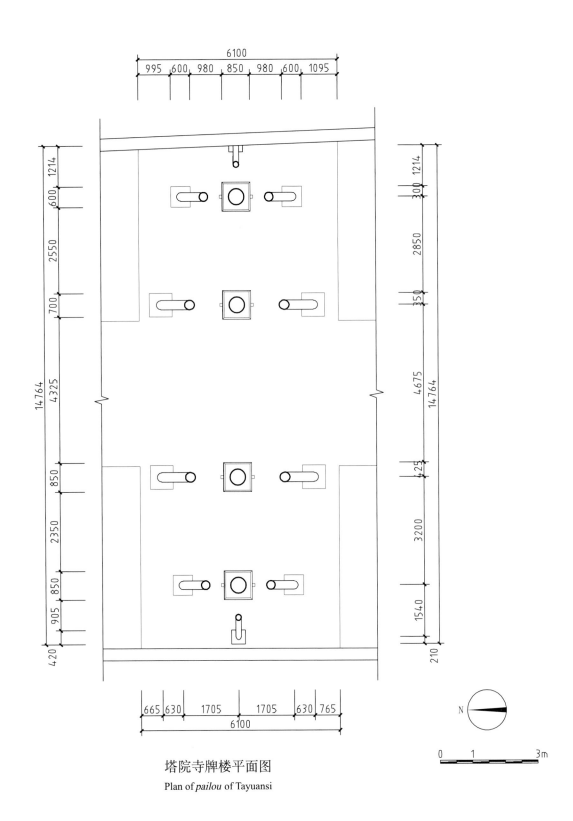

塔院寺牌楼平面图

Plan of *pailou* of Tayuansi

0 1 3m

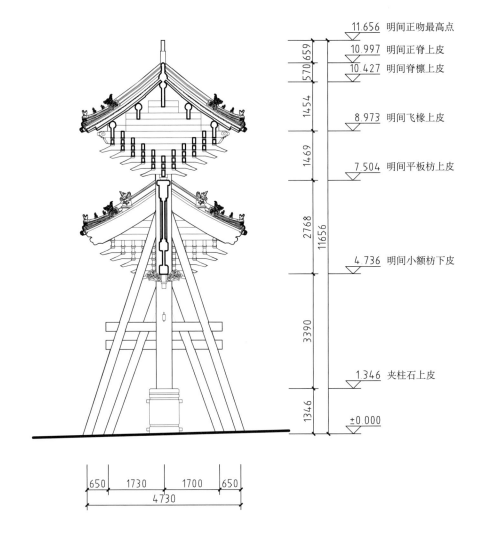

11 656 明间正吻最高点
10 997 明间正脊上皮
10 427 明间脊檩上皮

8 973 明间飞椽上皮

7 504 明间平板枋上皮

4 736 明间小额枋下皮

1 346 夹柱石上皮

±0.000

650 1730 1700 650
4730

塔院寺牌楼剖面图

Section of *pailou* of Tayuansi

151

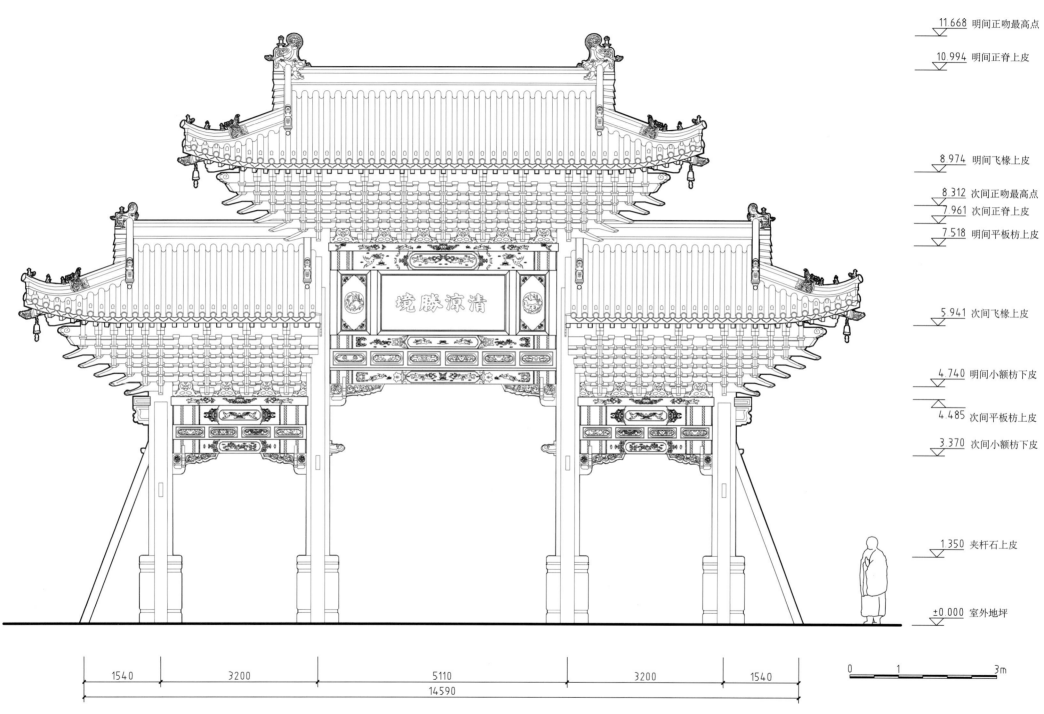

11.668 明间正吻最高点

10.994 明间正脊上皮

8.974 明间飞椽上皮

8.312 次间正吻最高点
7.961 次间正脊上皮
7.518 明间平板枋上皮

5.941 次间飞椽上皮

4.740 明间小额枋下皮

4.485 次间平板枋上皮

3.370 次间小额枋下皮

1.350 夹杆石上皮

±0.000 室外地坪

| 1540 | 3200 | 5110 | 3200 | 1540 |

14590

0　1　　3m

塔院寺牌楼正立面图

Front elevation of *pailou* of Tayuansi

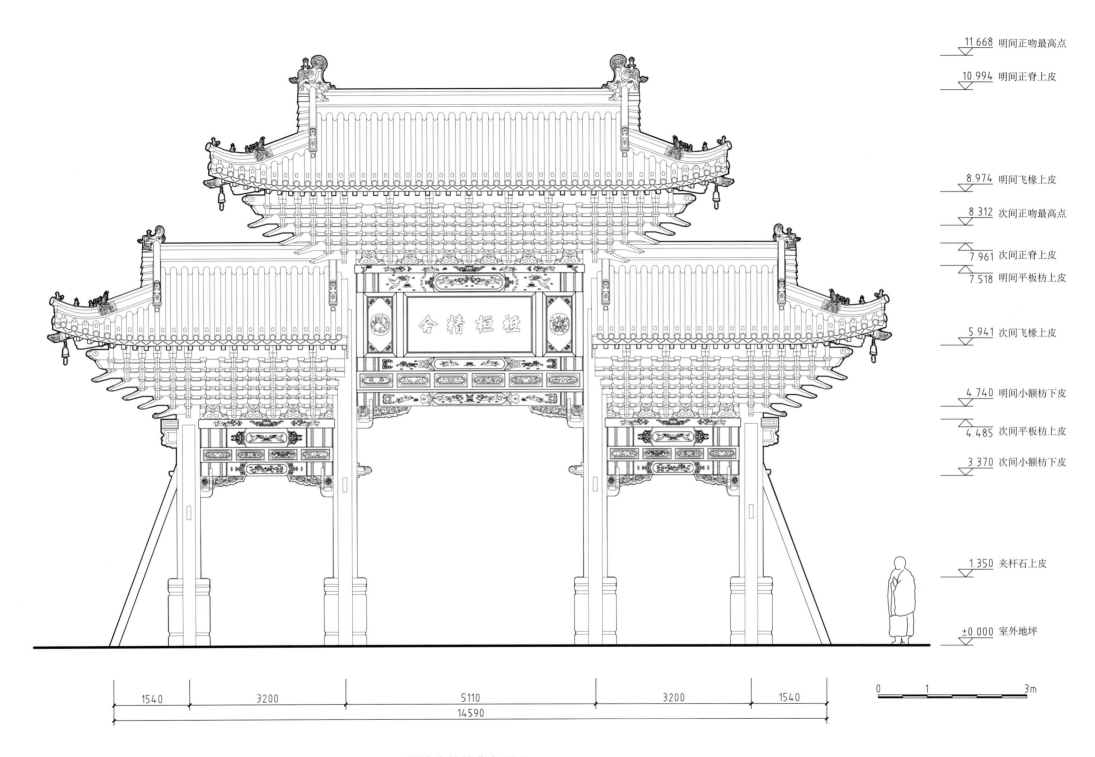

11.668 明间正吻最高点

10.994 明间正脊上皮

8.974 明间飞椽上皮

8.312 次间正吻最高点

7.961 次间正脊上皮
7.518 明间平板枋上皮

5.941 次间飞椽上皮

4.740 明间小额枋下皮

4.485 次间平板枋上皮

3.370 次间小额枋下皮

1.350 夹杆石上皮

±0.000 室外地坪

0   1   3m

1540   3200   5110   3200   1540
14590

塔院寺牌楼背立面图
Rear elevation of *pailou* of Tayuansi

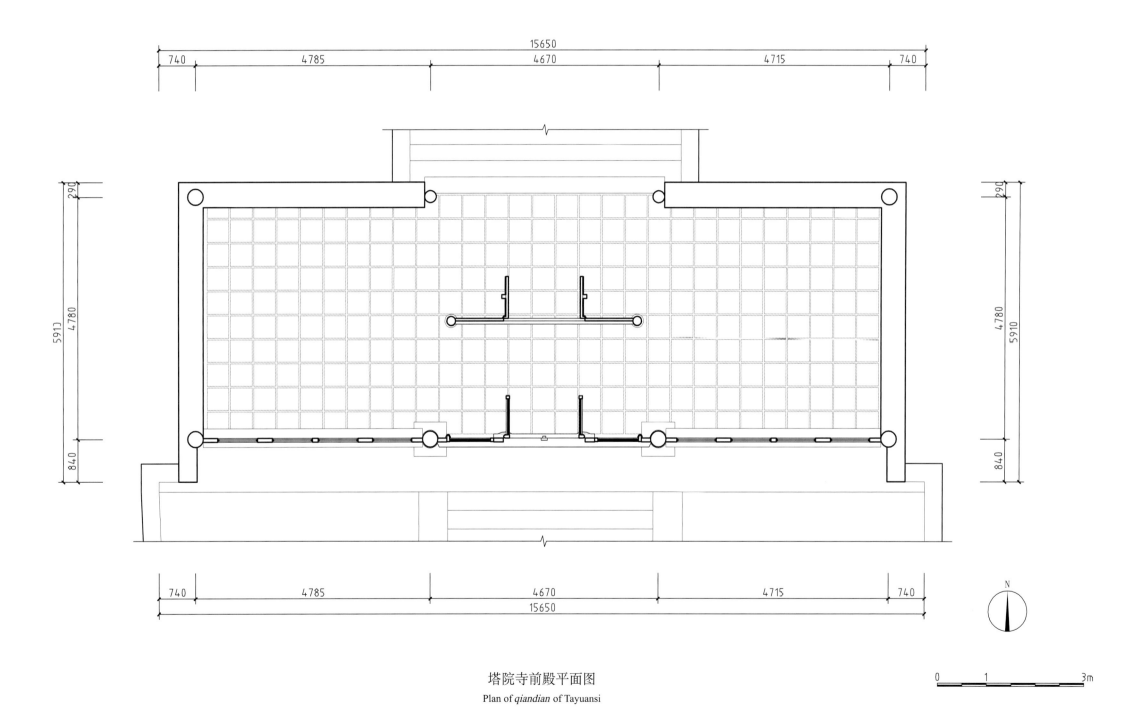

塔院寺前殿平面图

Plan of *qiandian* of Tayuansi

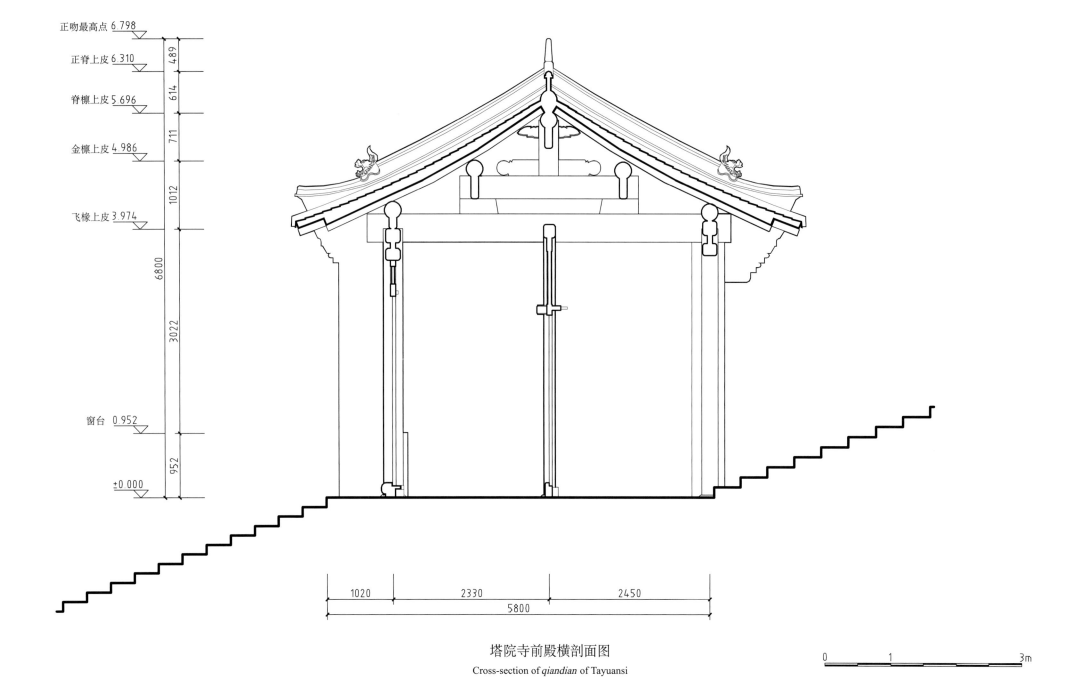

正吻最高点 6.798

正脊上皮 6.310

脊檩上皮 5.696

金檩上皮 4.986

飞椽上皮 3.974

窗台 0.952

±0.000

489

614

711

1012

6800

3022

952

1020 2330 2450

5800

塔院寺前殿横剖面图

Cross-section of *qiandian* of Tayuansi

0   1   3m

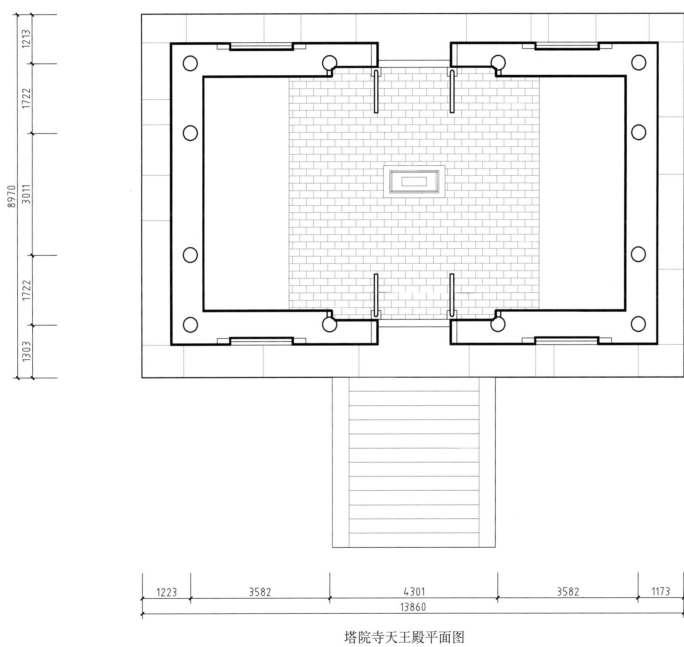

1213
1722
3011
8970
1722
1303

1213
1722
3011
8970
1722
1303

1223 3582 4301 3582 1173
13860

塔院寺天王殿平面图
Plan of Tianwangdian of Tayuansi

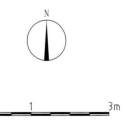

N

0 1 3m

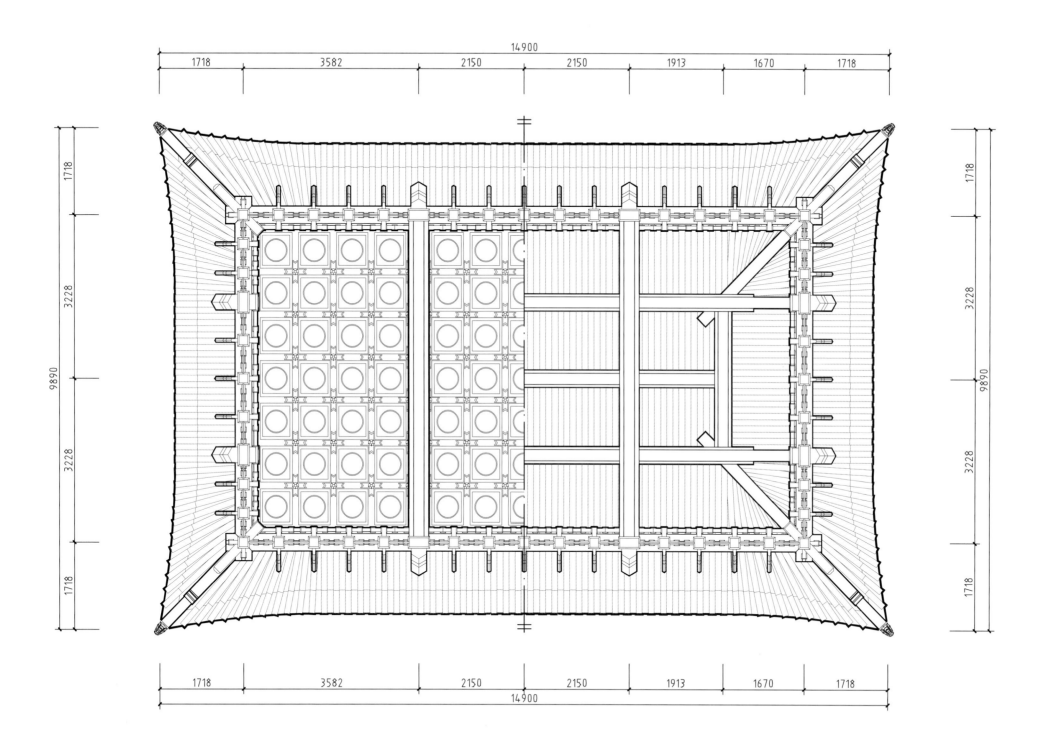

塔院寺天王殿梁架仰视平面图

Plan of framework of Tianwangdian of Tayuansi as seen from below

0 1 3m

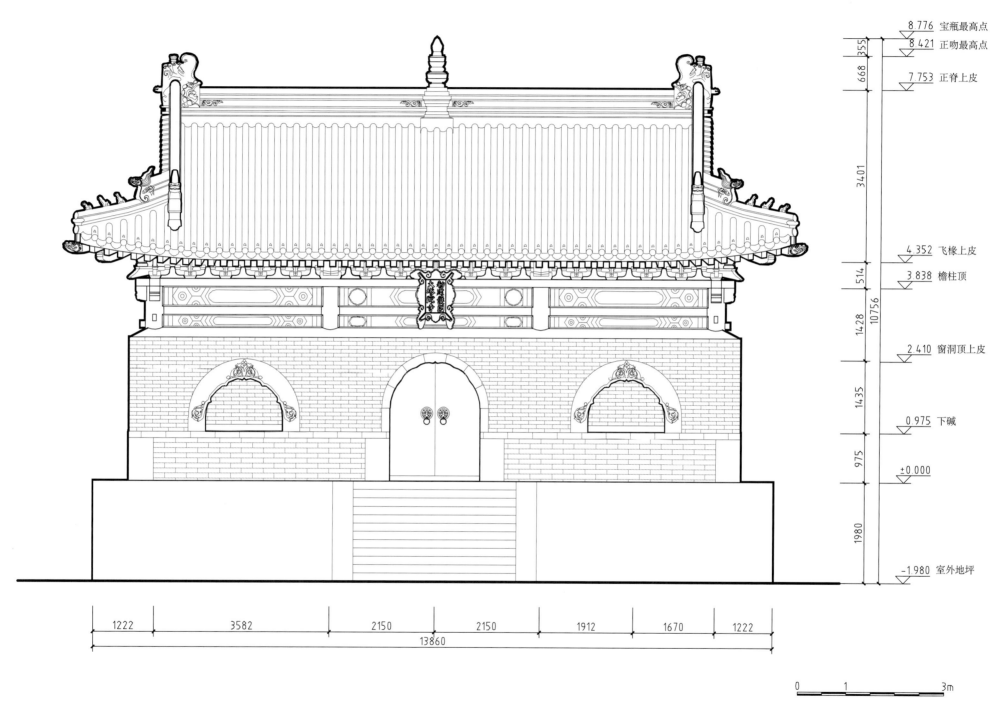

8 776 宝瓶最高点
8 421 正吻最高点
7 753 正脊上皮

4 352 飞椽上皮
3 838 檐柱顶

2 410 窗洞顶上皮

0 975 下碱

±0.000

-1 980 室外地坪

355
668
3401
514
1428
10756
1435
975
1980

1222　3582　2150　2150　1912　1670　1222
13860

0　1　3m

塔院寺天王殿正立面图
Front elevation of Tianwangdian of Tayuansi

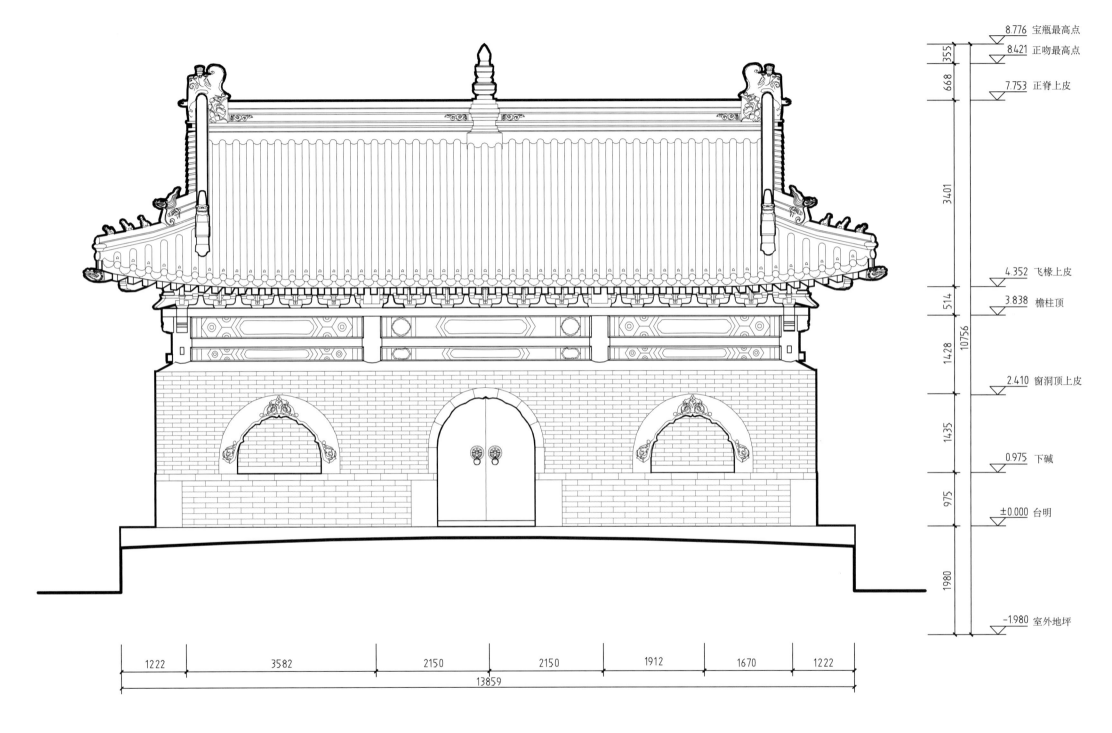

8.776 宝瓶最高点
8.421 正吻最高点
7.753 正脊上皮
4.352 飞椽上皮
3.838 檐柱顶
2.410 窗洞顶上皮
0.975 下碱
±0.000 台明
-1.980 室外地坪

355
668
3401
514
1428
1435
975
1980
10756

1222　3582　2150　2150　1912　1670　1222
13859

塔院寺天王殿背立面图
Rear elevation of Tianwangdian of Tayuansi

0　1　3m

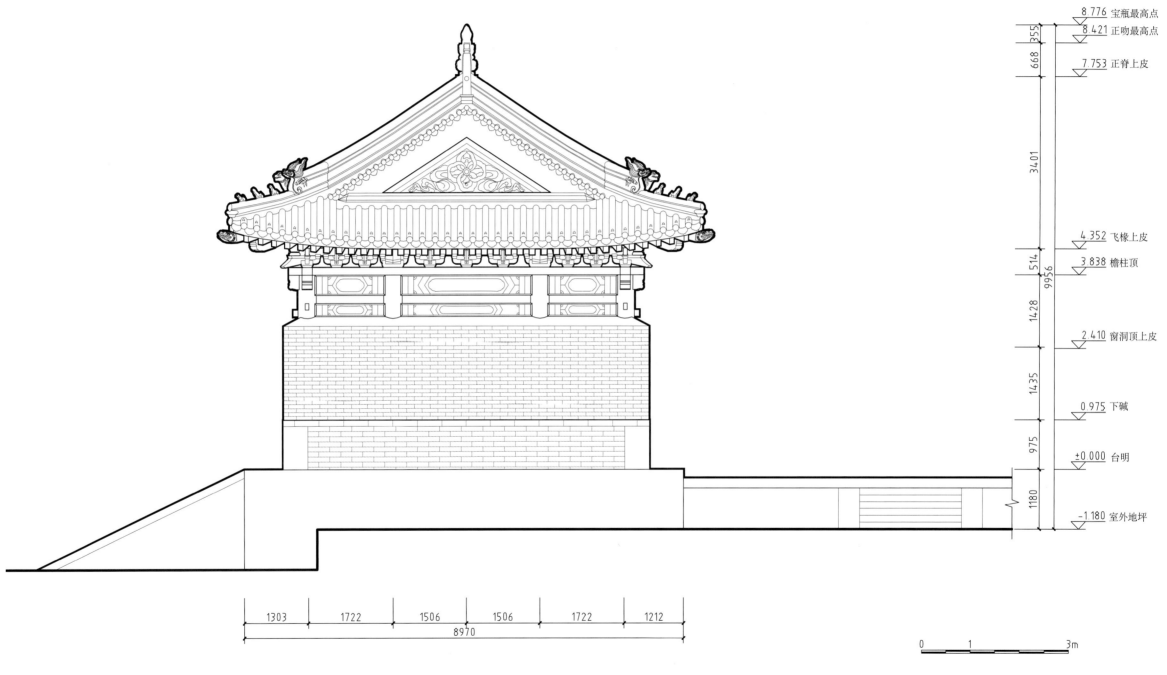

8.776 宝瓶最高点
8.421 正吻最高点
7.753 正脊上皮
4.352 飞椽上皮
3.838 檐柱顶
2.410 窗洞顶上皮
0.975 下碱
±0.000 台明
-1.180 室外地坪

355
668
3401
514
9956
1428
1435
975
1180

1303 1722 1506 1506 1722 1212
8970

0 1 3m

塔院寺天王殿侧立面图

Side elevation of Tianwangdian of Tayuansi

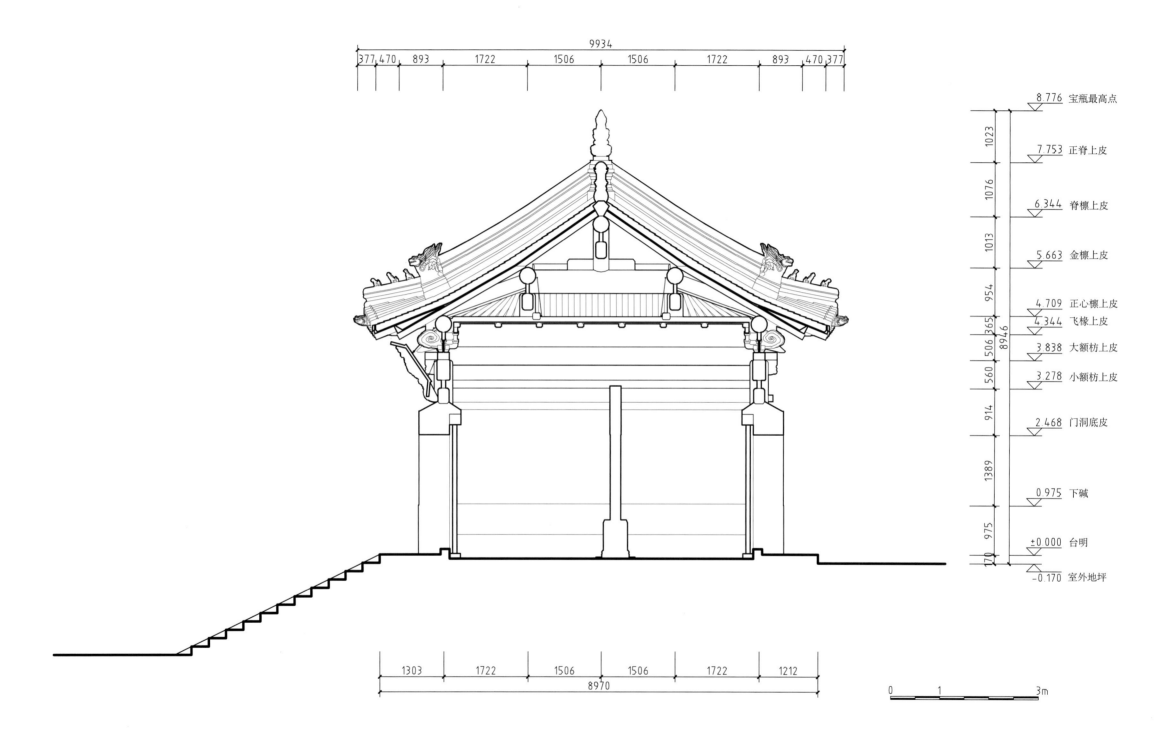

9934

377 470 893 1722 1506 1506 1722 893 470 377

8.776 宝瓶最高点
1023
7.753 正脊上皮
1076
6.344 脊檩上皮
1013
5.663 金檩上皮
954
4.709 正心檩上皮
365
4.344 飞椽上皮
506
3.838 大额枋上皮
560
3.278 小额枋上皮
914
2.468 门洞底皮
1389
0.975 下碱
975
±0.000 台明
170
-0.170 室外地坪
8946

1303 1722 1506 1506 1722 1212
8970

0 1 3m

塔院寺天王殿横剖面图
Cross-section of Tianwangdian of Tayuansi

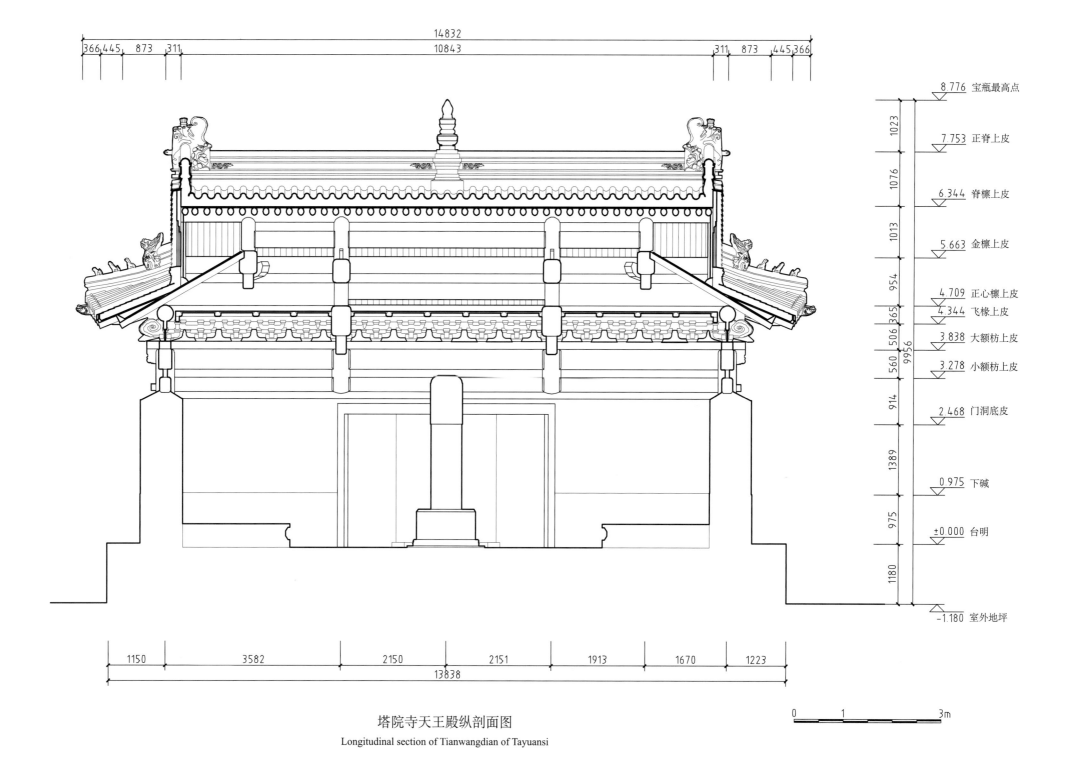

14832
10843
366 445 873 311
311 873 445 366

8.776 宝瓶最高点
1023
7.753 正脊上皮
1076
6.344 脊檩上皮
1013
5.663 金檩上皮
954
4.709 正心檩上皮
365
4.344 飞椽上皮
506
3.838 大额枋上皮
9956
560
3.278 小额枋上皮
914
2.468 门洞底皮
1389
0.975 下碱
975
±0.000 台明
1180
-1.180 室外地坪

1150 3582 2150 2151 1913 1670 1223
13838

塔院寺天王殿纵剖面图
Longitudinal section of Tianwangdian of Tayuansi

0 1 3m

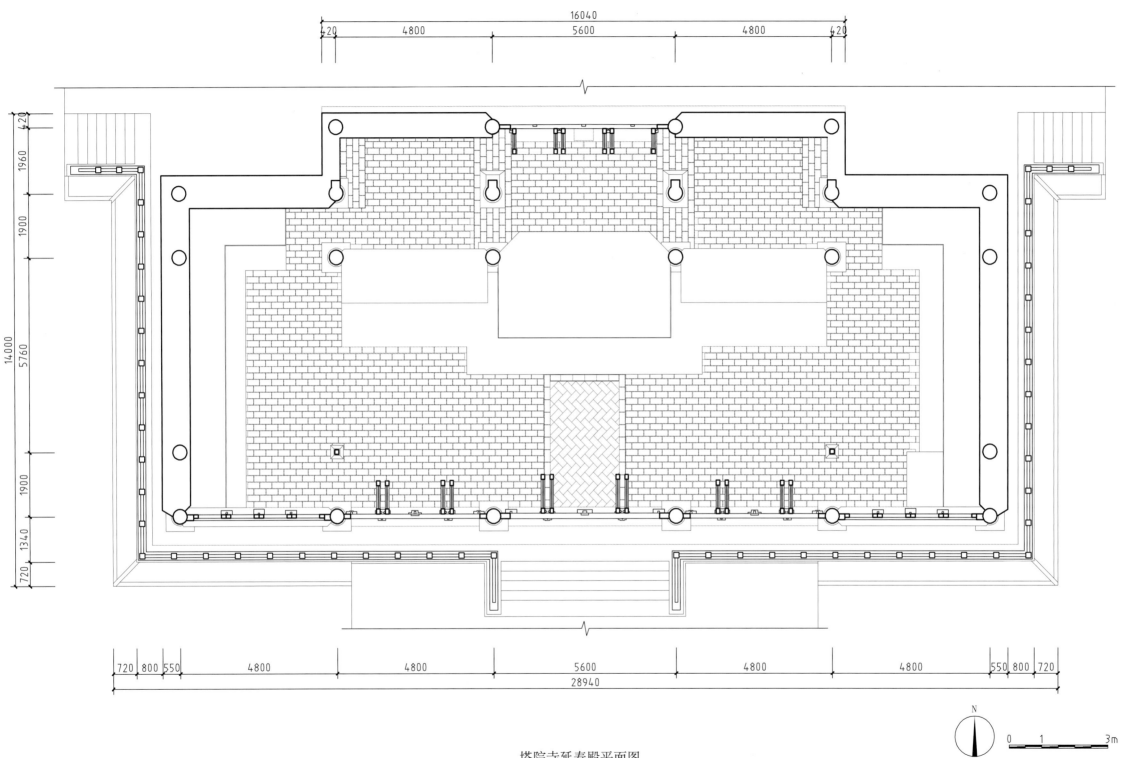

塔院寺延寿殿平面图
Plan of Yanshoudian of Tayuansi

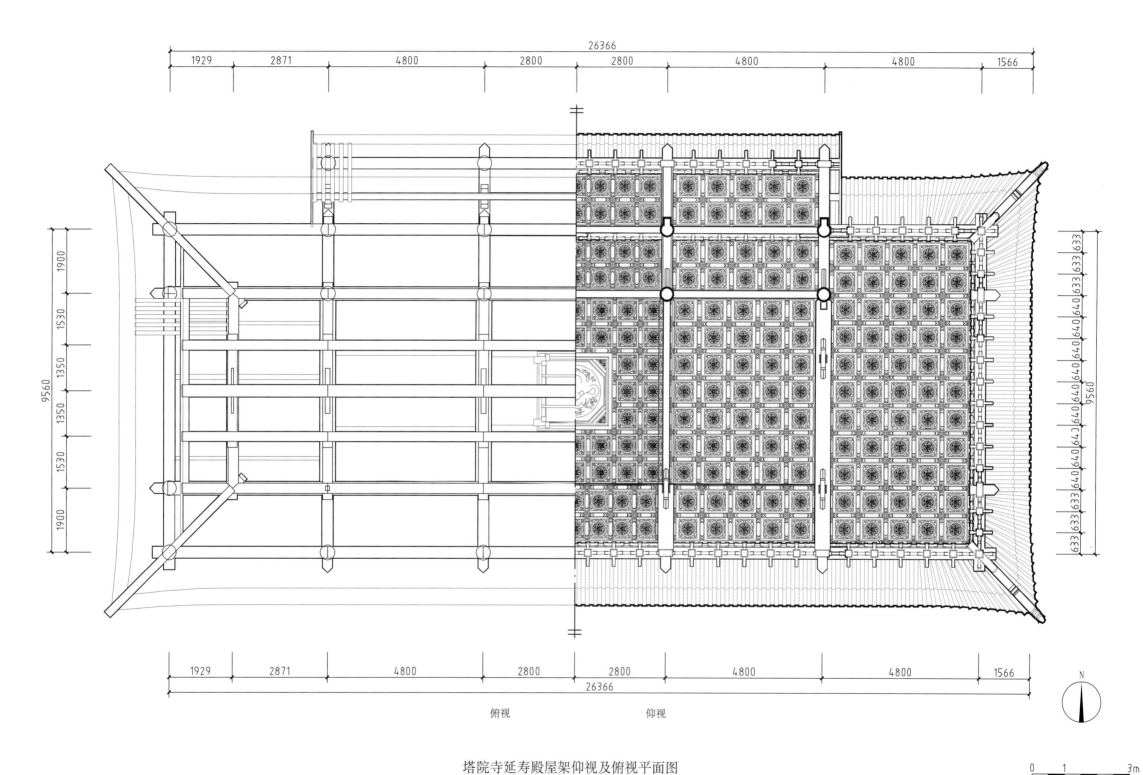

俯视　　　　　　　仰视

塔院寺延寿殿屋架仰视及俯视平面图

Plan of roof framework of Yanshoudian of Tayuansi as seen from below and above

0　1　　　3m

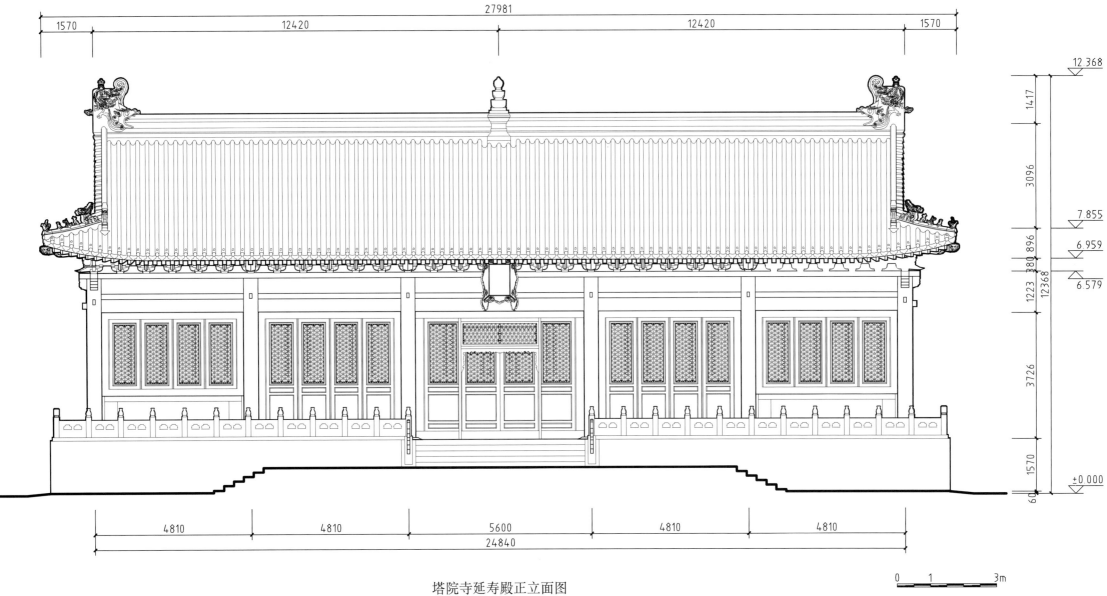

27981

1570   12420   12420   1570

12 368

1417

3096

7 855

896
380
1223

6 959

6 579

12368

3726

1570

60

±0 000

4810   4810   5600   4810   4810

24840

塔院寺延寿殿正立面图

Front elevation of Yanshoudian of Tayuansi

0   1   3m

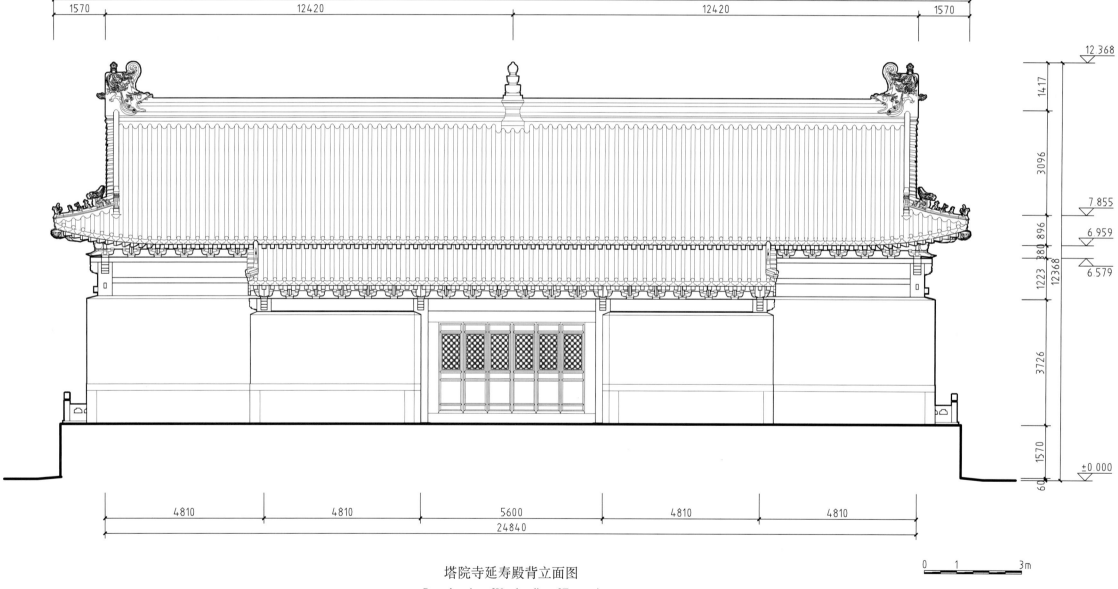

塔院寺延寿殿背立面图

Rear elevation of Yanshoudian of Tayuansi

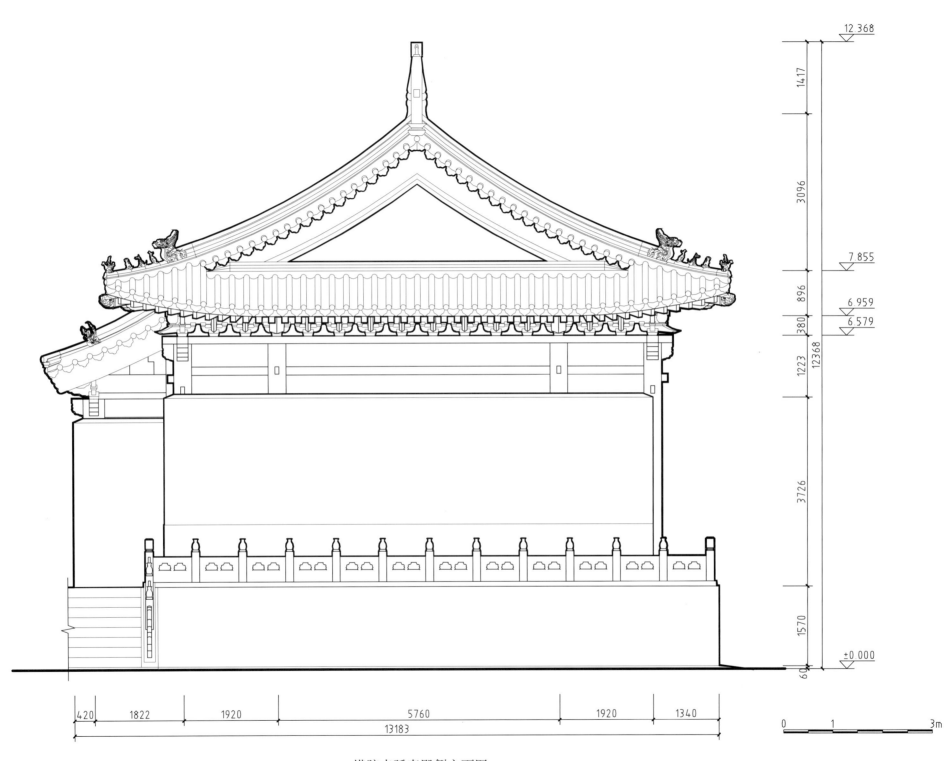

塔院寺延寿殿侧立面图

Side elevation of Yanshoudian of Tayuansi

167

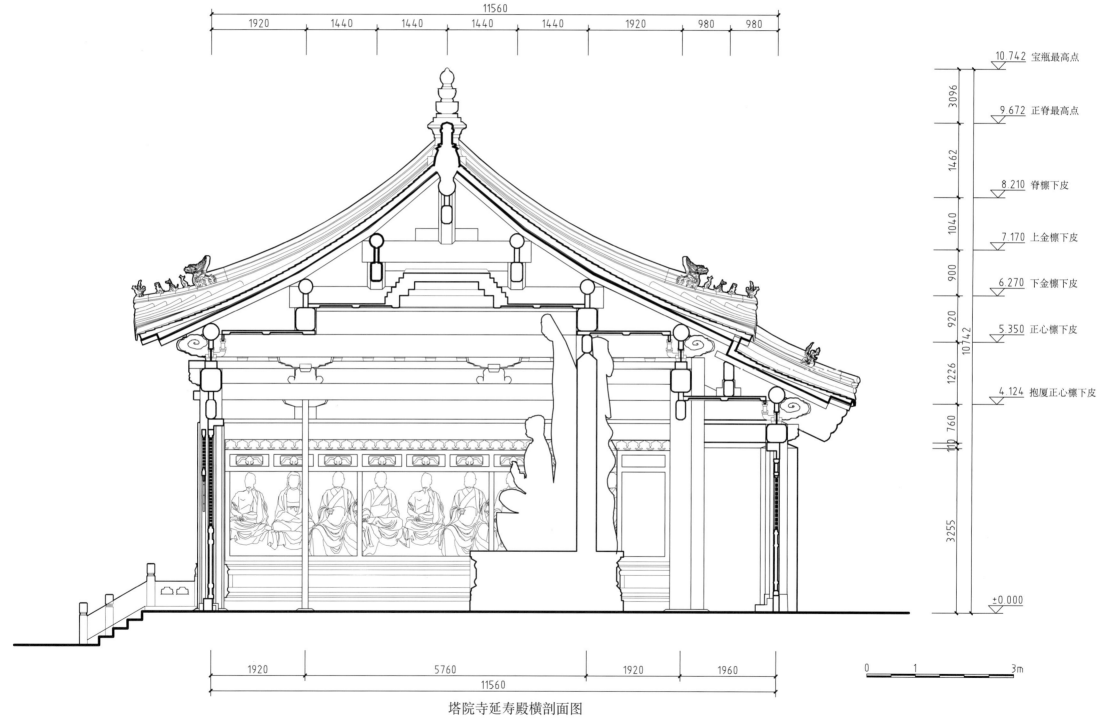

11560

1920 1440 1440 1440 1440 1920 980 980

3096 10.742 宝瓶最高点

1462 9.672 正脊最高点

1040 8.210 脊檩下皮

900 7.170 上金檩下皮

920 6.270 下金檩下皮

1226 5.350 正心檩下皮

10742

760 4.124 抱厦正心檩下皮

110

3255 ±0.000

1920 5760 1920 1960
11560

0 1 3m

塔院寺延寿殿横剖面图
Cross-section of Yanshoudian of Tayuansi

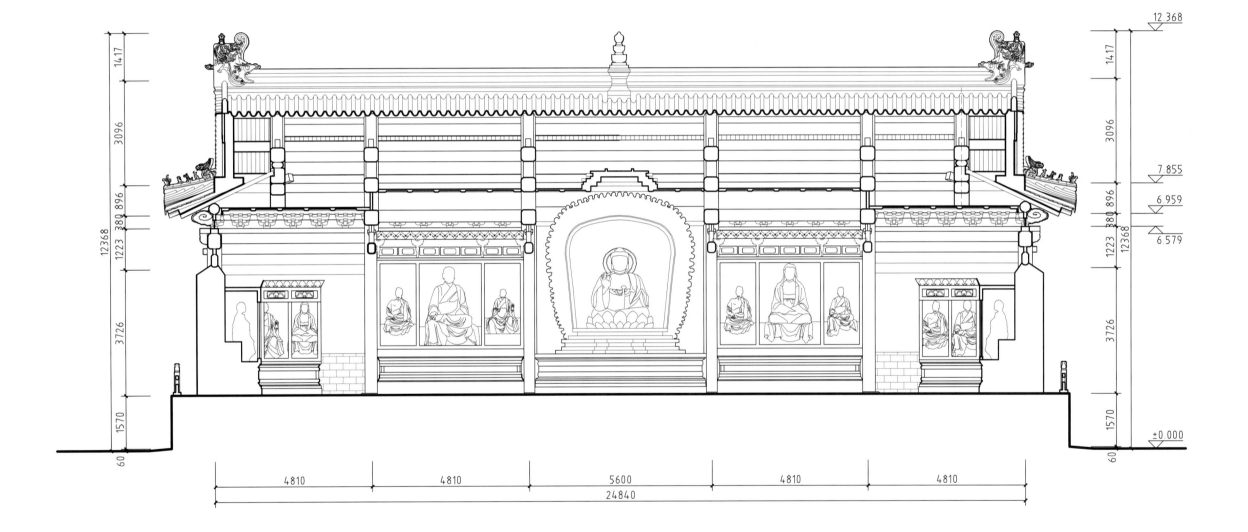

塔院寺延寿殿纵剖面图

Longitudinal section of Yanshoudian of Tayuansi

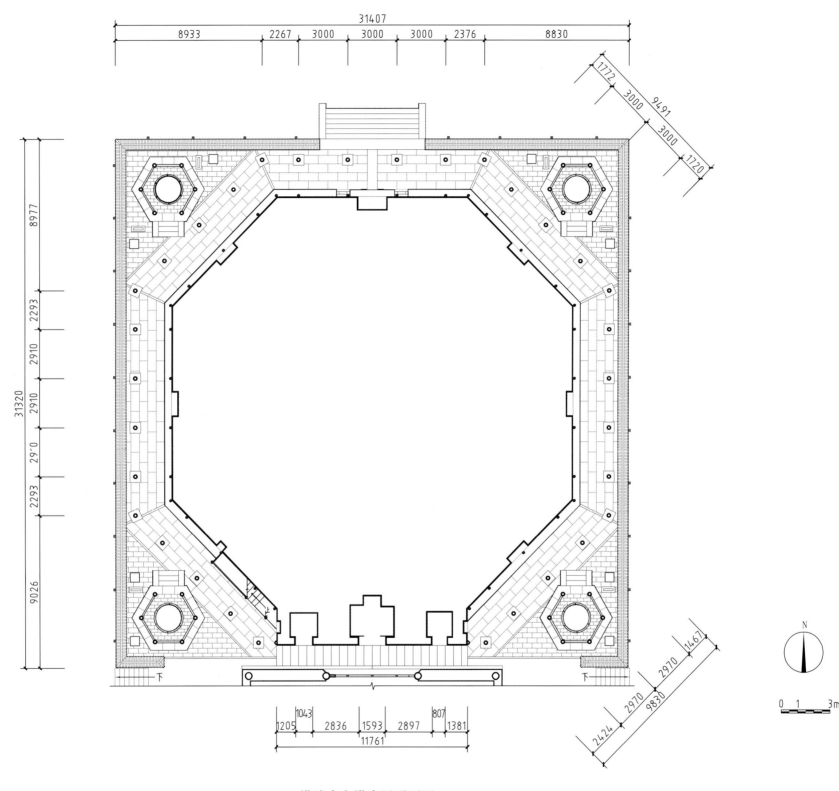

塔院寺白塔底层平面图

Plan of basement of Baita of Tayuansi

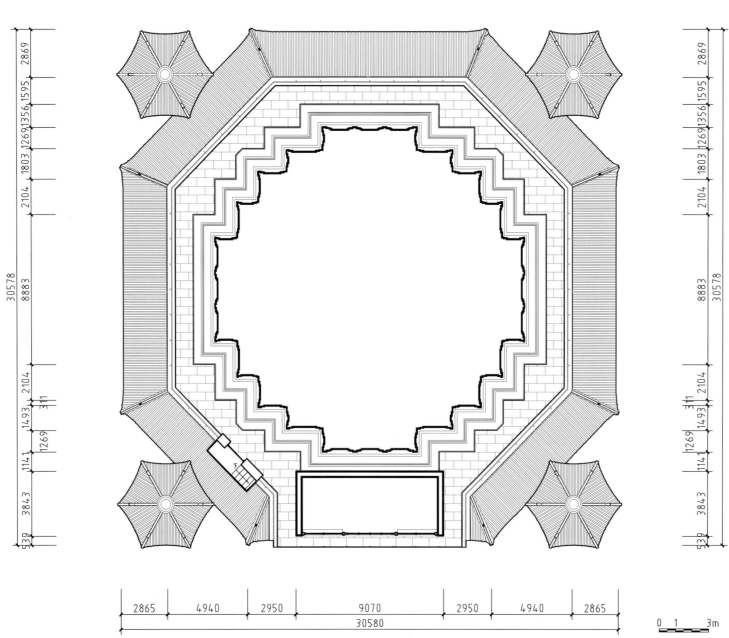

塔院寺白塔二层平面图
Plan of second floor of Baita of Tayuansi

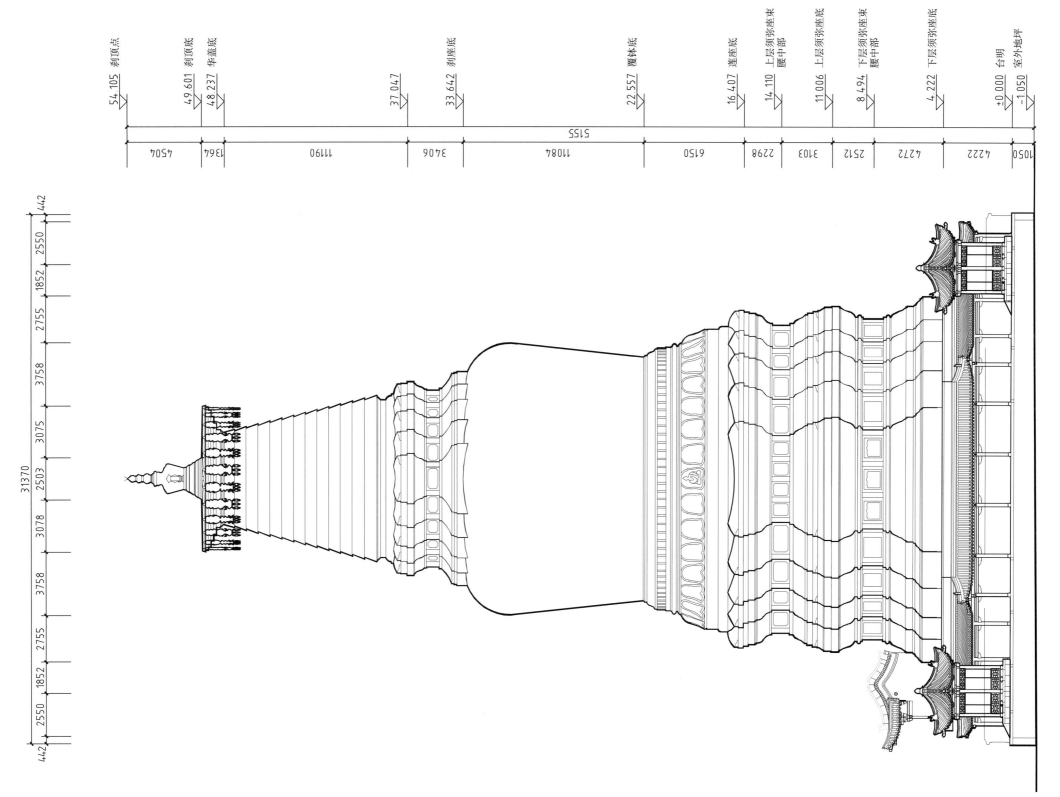

54.105 刹顶点

49.601 刹顶座底
48.237 华盖底

37.047 刹座底

33.642 刹座底

22.557 覆钵底

16.407 莲座底

14.110 上层须弥座束腰中部

11.006 上层须弥座底

8.494 下层须弥座束腰中部

4.222 下层须弥座底

±0.000 台明
-1.050 室外地坪

4504
1364
11190
3406
11084
6150
2298
3103
2512
4272
4222
1050

5155

442
2550
1852
2755
3758
3075
2503
3078
3758
2755
1852
2550
442

31370

0 1   3 m

塔院寺白塔东立面图

East elevation of Baita of Tayuansi

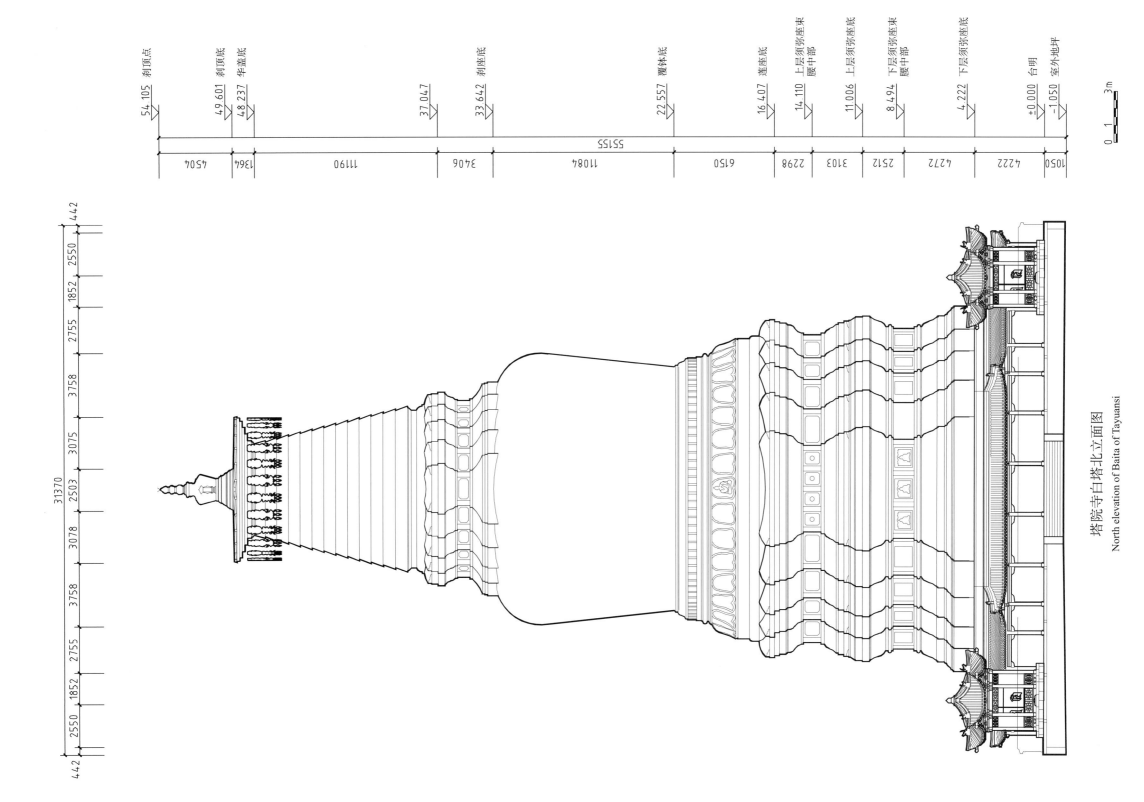

54.105 刹顶点

49.601 刹顶底
48.237 华盖底

37.047

33.642 刹座底

22.557 覆钵底

16.407 莲座底
14.110 上层须弥座束腰中部
11.006 上层须弥座底
8.494 下层须弥座束腰中部
4.222 下层须弥座底

±0.000 台明
-1.050 室外地坪

4504
1364
11190
3406
11084
6150
2298
3103
2512
4.272
4.222
1050

55155

0 1 3m

2550
1852
2755
3758
3075
2503
3078
3758
2755
1852
2550

31370

442
442

塔院寺白塔北立面图
North elevation of Baita of Tayuansi

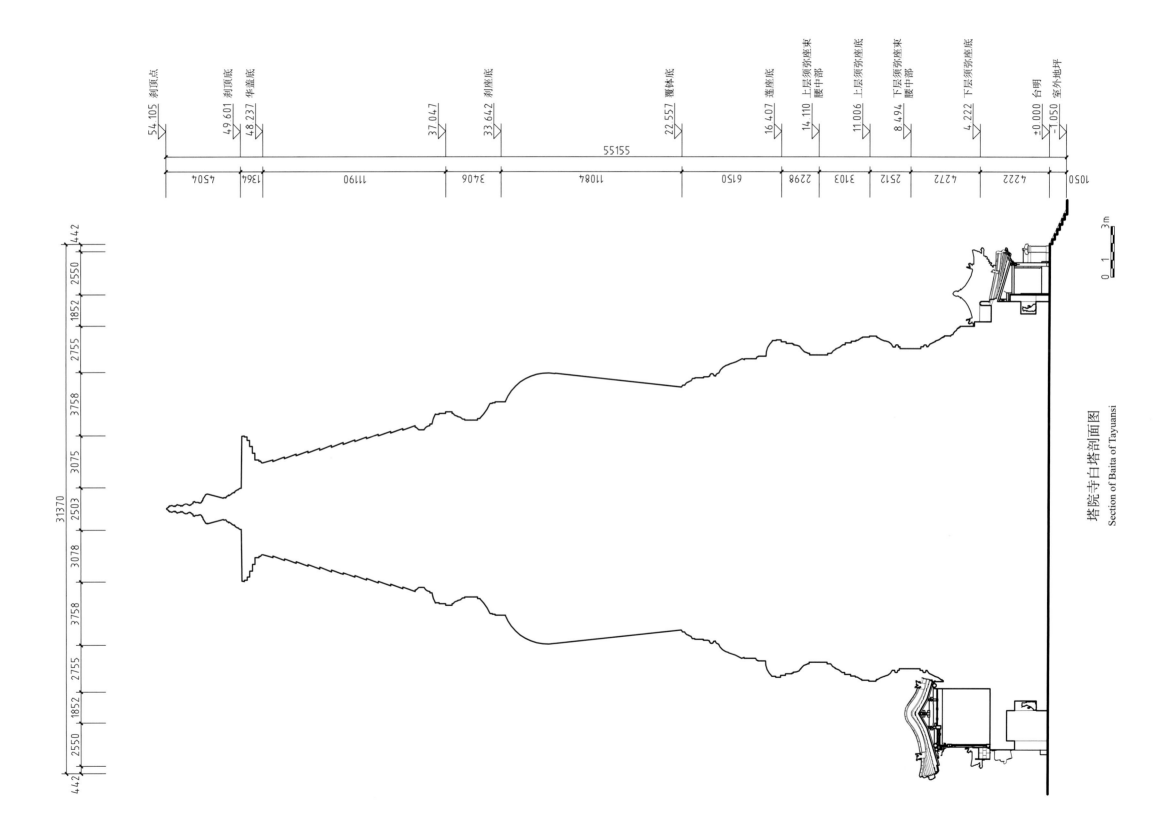

塔院寺白塔剖面图
Section of Baita of Tayuansi

54.105 刹顶点
49.601 刹顶底
48.237 华盖底
37.047 刹座底
33.642 刹座座底
22.557 覆钵底
16.407 莲座底
14.110 上层须弥座束腰中部
11.006 上层须弥座底
8.494 下层须弥座束腰中部
4.222 下层须弥座底
±0.000 台明
-1.050 室外地坪

55155

4504
1364
11190
3406
11084
6150
2298
3103
2512
4272
4222
1050

442 2550 1852 2755 3758 3075 2503 3078 3758 2755 1852 2550 442

31370

0 1 3m

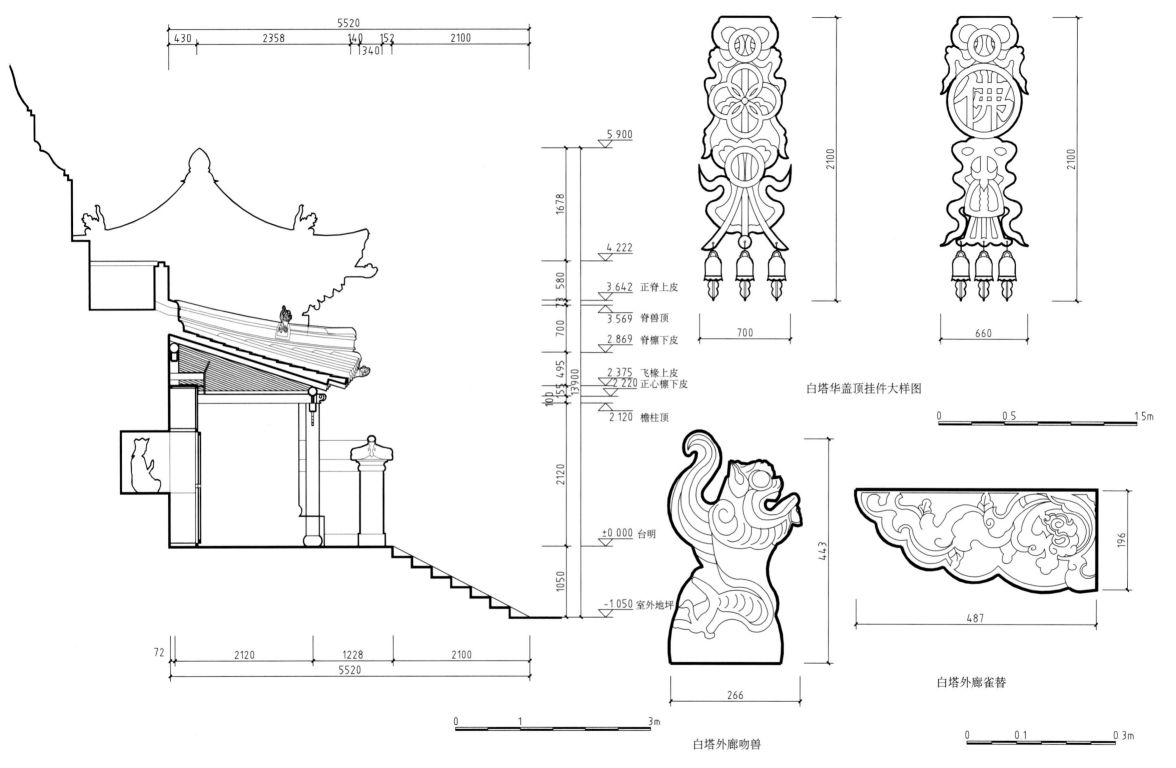

5520
430　2358　140　152　2100
1340

2100

2100

700

660

白塔华盖顶挂件大样图

5.900

1678

4.222

580　3.642　正脊上皮
3.569　脊兽顶
700　2.869　脊檩下皮
495　2.375　飞椽上皮
2.220　正心檩下皮
100　2.120　檐柱顶
155
13900

2120

±0.000　台明

1050

-1.050　室外地坪

443

266

白塔外廊吻兽

196

487

白塔外廊雀替

0　0.5　15m

0　0.1　0.3m

72　2120　1228　2100
5520

0　1　3m

塔院寺白塔底层外廊剖面图
Section of basement corridor of Baita of Tayuansi

塔院寺白塔构件大样图
Structural components of Baita of Tayuansi

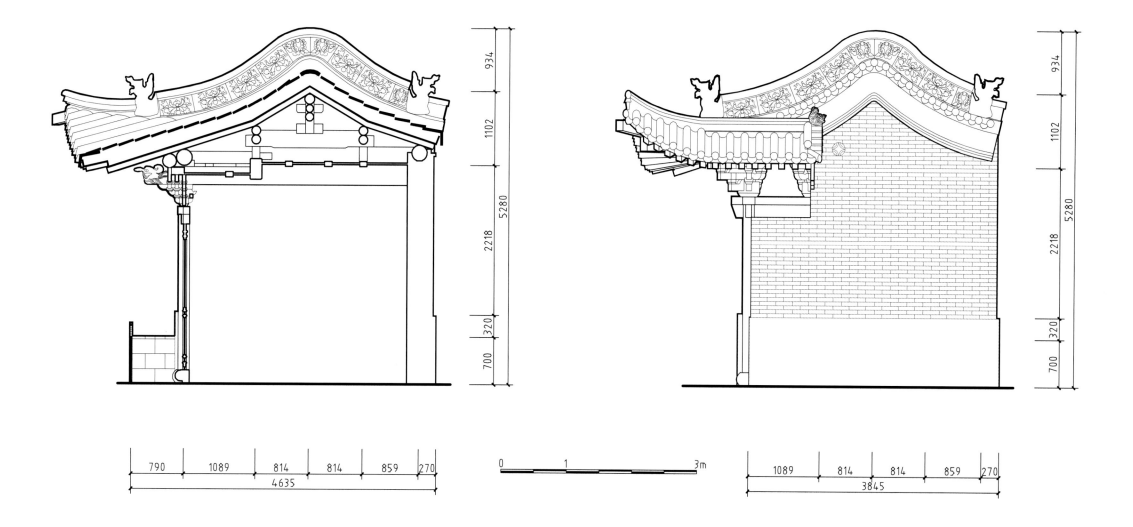

934 1102 2218 320 700 5280

790 1089 814 814 859 270
4635

0　　　1　　　3m

1089 814 814 859 270
3845

934 1102 2218 320 700 5280

塔院寺白塔献殿剖面图
Section of *xiandian* of Baita of Tayuansi

塔院寺白塔献殿侧立面图
Side elevation of *xiandian* of Baita of Tayuansi

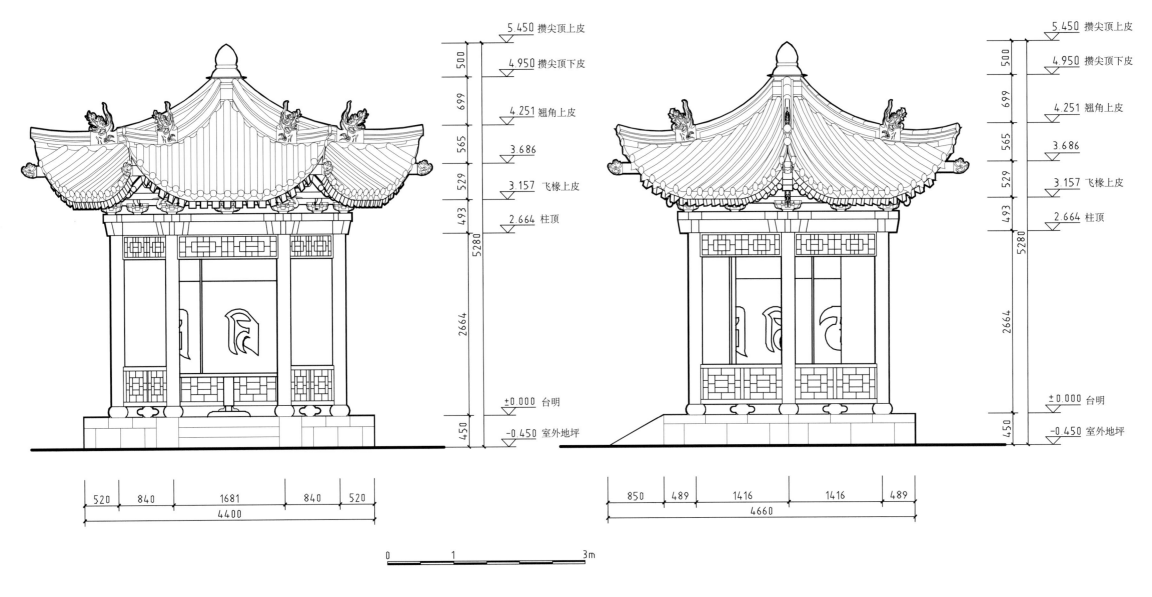

5.450 攒尖顶上皮
500
4.950 攒尖顶下皮
699
4.251 翘角上皮
565
3.686
529
3.157 飞椽上皮
493
2.664 柱顶
5280
2664
±0.000 台明
450
-0.450 室外地坪

5.450 攒尖顶上皮
500
4.950 攒尖顶下皮
699
4.251 翘角上皮
565
3.686
529
3.157 飞椽上皮
493
2.664 柱顶
5280
2664
±0.000 台明
450
-0.450 室外地坪

520 840 1681 840 520
4400

850 489 1416 1416 489
4660

0 1 3m

塔院寺白塔六角亭正立面图
Front elevation of Liujiaoting of Baita of Tayuansi

塔院寺白塔六角亭侧立面图
Side elevation Liujiaoting of Baita of Tayuansi

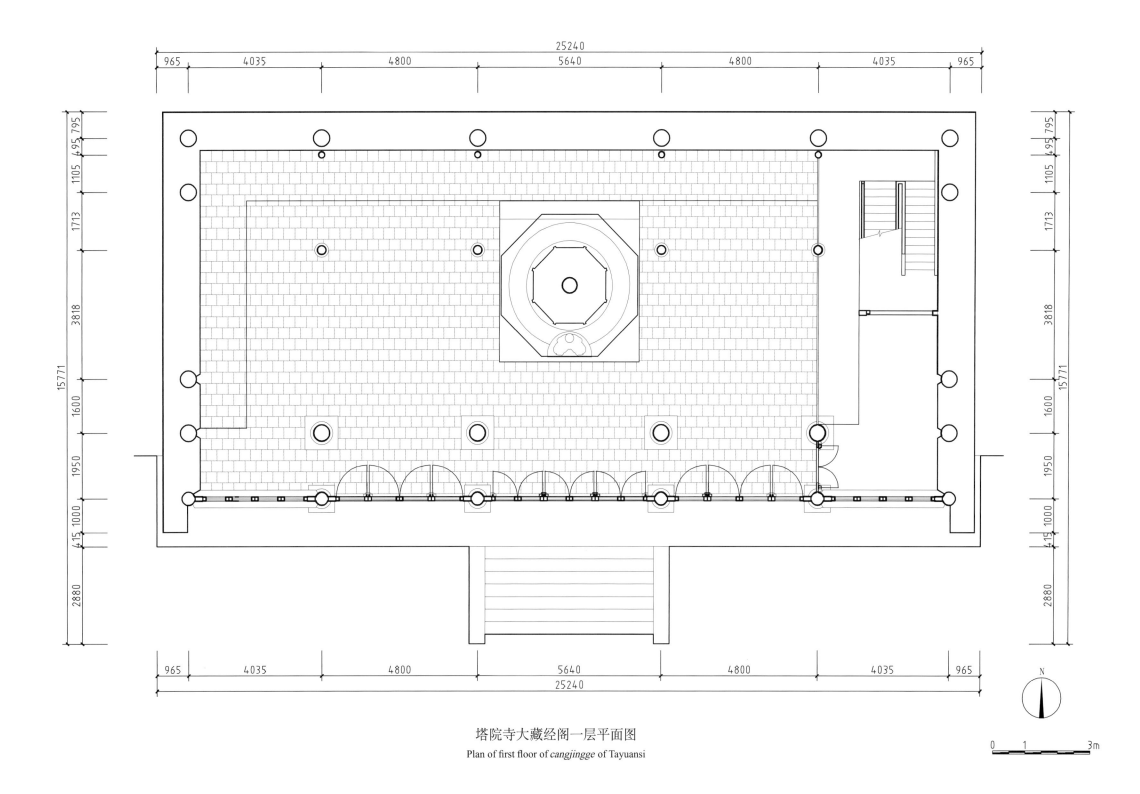

塔院寺大藏经阁一层平面图

Plan of first floor of *cangjingge* of Tayuansi

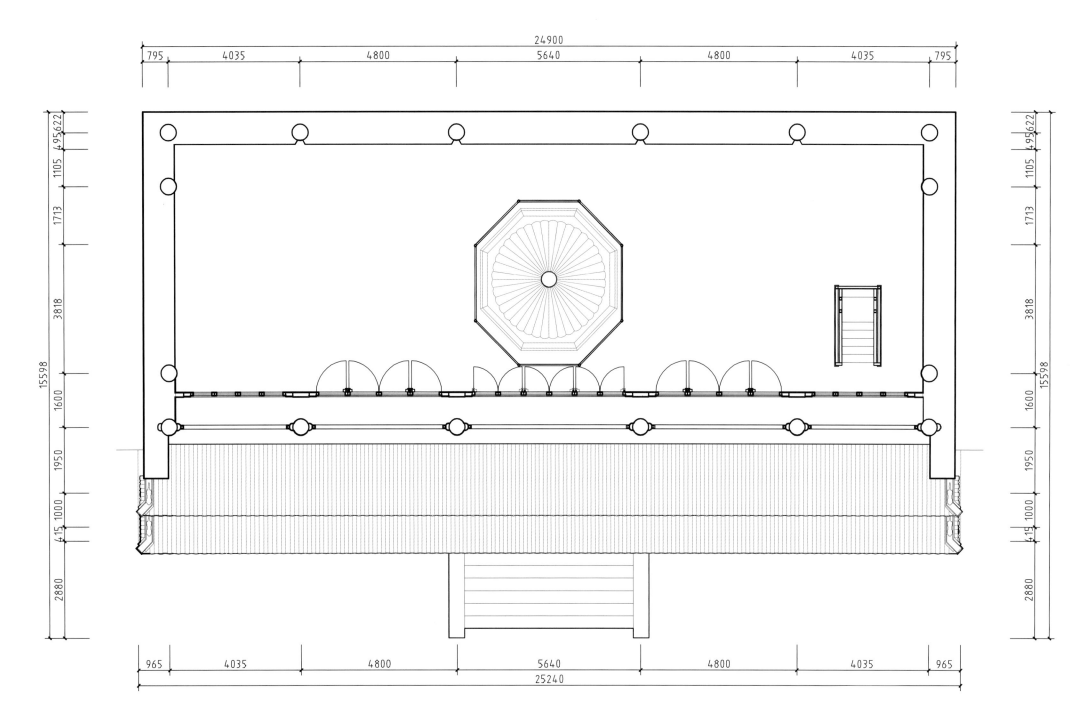

塔院寺大藏经阁二层平面图

Plan of second floor of *cangjingge* of Tayuansi

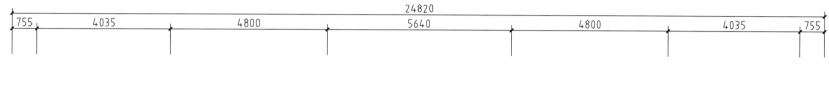

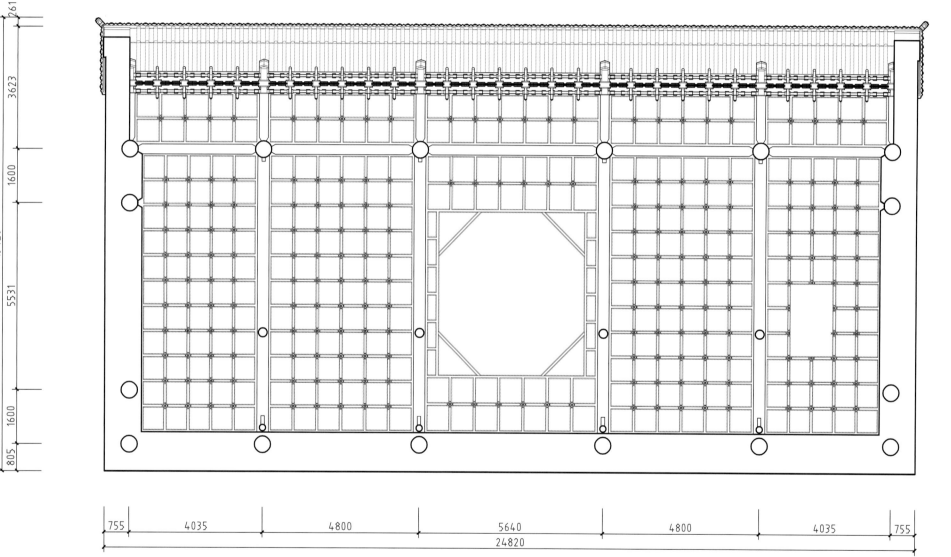

塔院寺大藏经阁一层仰视平面图

Plan of first floor of *cangjingge* of Tayuansi as seen from below

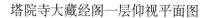

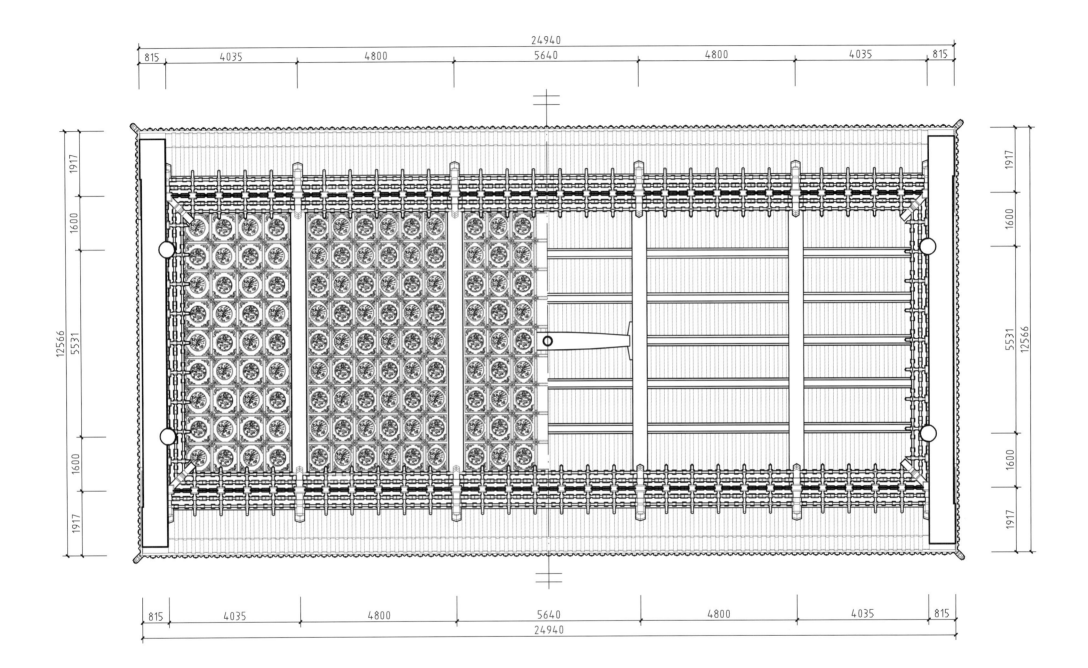

塔院寺大藏经阁二层仰视平面图

Plan of second floor of *cangjingge* of Tayuansi as seen from below

0   1   3m

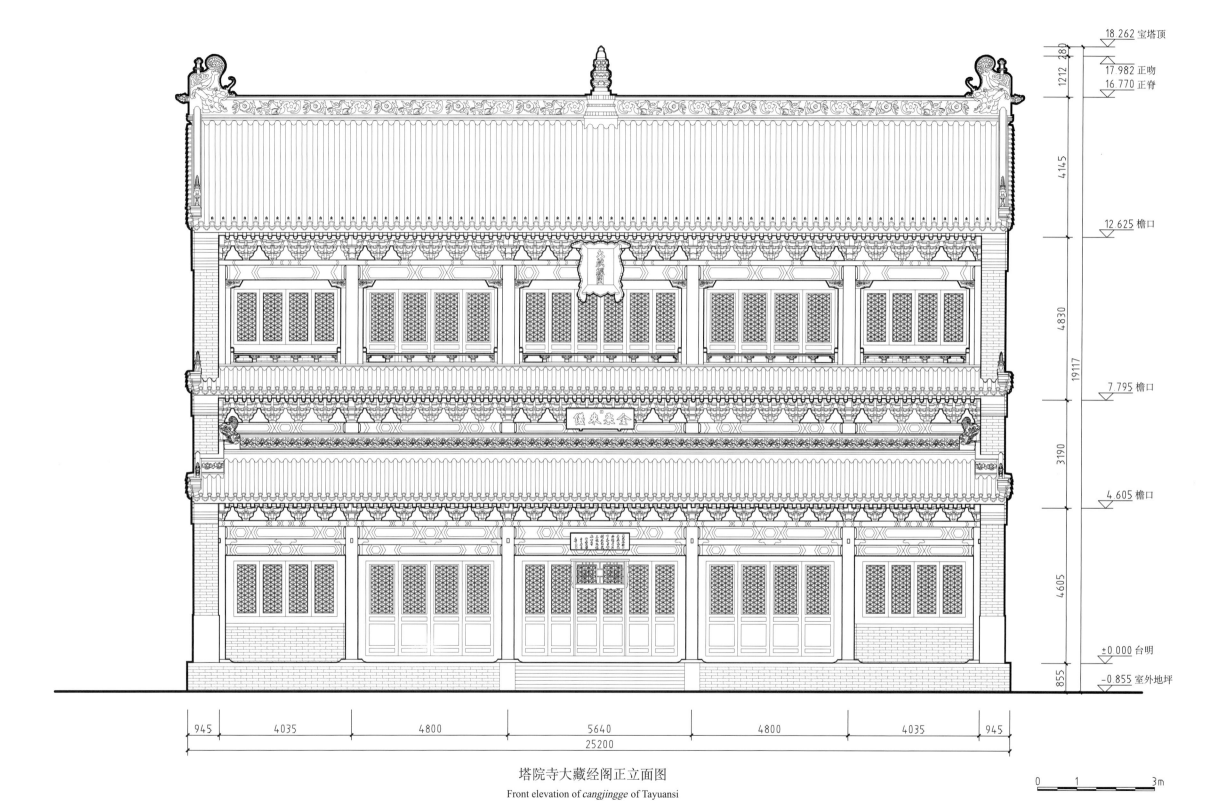

18.262 宝塔顶
17.982 正吻
16.770 正脊
12.625 檐口
7.795 檐口
4.605 檐口
±0.000 台明
-0.855 室外地坪

280
1212
4145
4830
19117
3190
4605
855

945 4035 4800 5640 4800 4035 945
25200

塔院寺大藏经阁正立面图
Front elevation of *cangjingge* of Tayuansi

0   1   3m

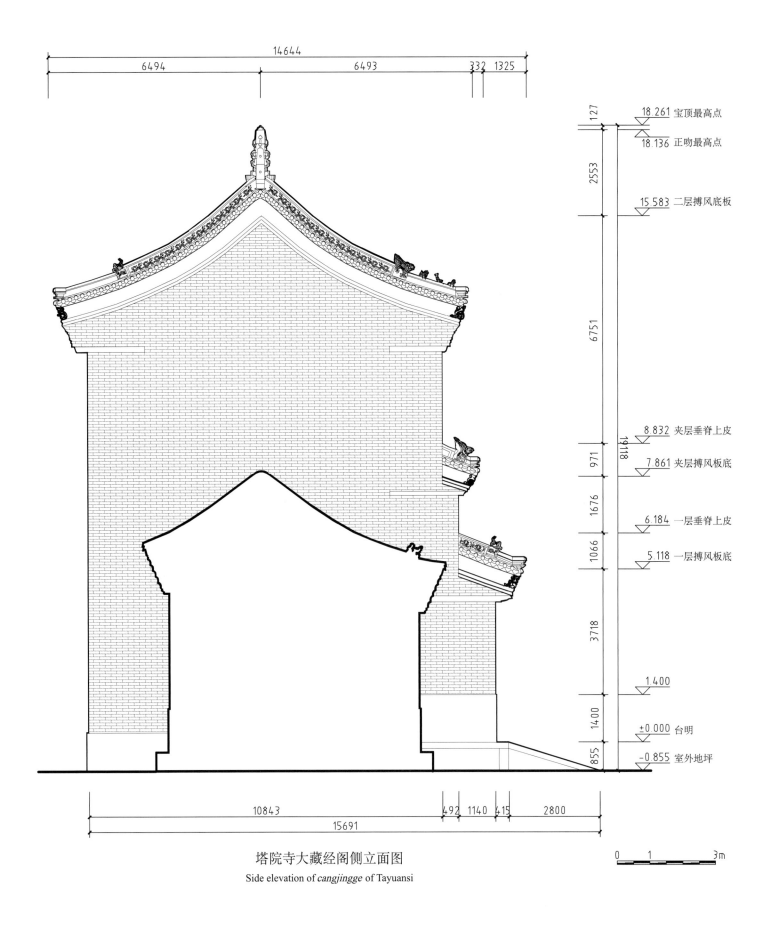

14644
6494   6493   332 1325

127   18.261 宝顶最高点
2553   18.136 正吻最高点

15.583 二层搏风底板

6751

19118

8.832 夹层垂脊上皮
971
7.861 夹层搏风板底

1676
6.184 一层垂脊上皮

1066
5.118 一层搏风板底

3718

1.400

1400
±0.000 台明

855   -0.855 室外地坪

10843   492 1140 415   2800
15691

塔院寺大藏经阁侧立面图
Side elevation of *cangjingge* of Tayuansi

0   1   3m

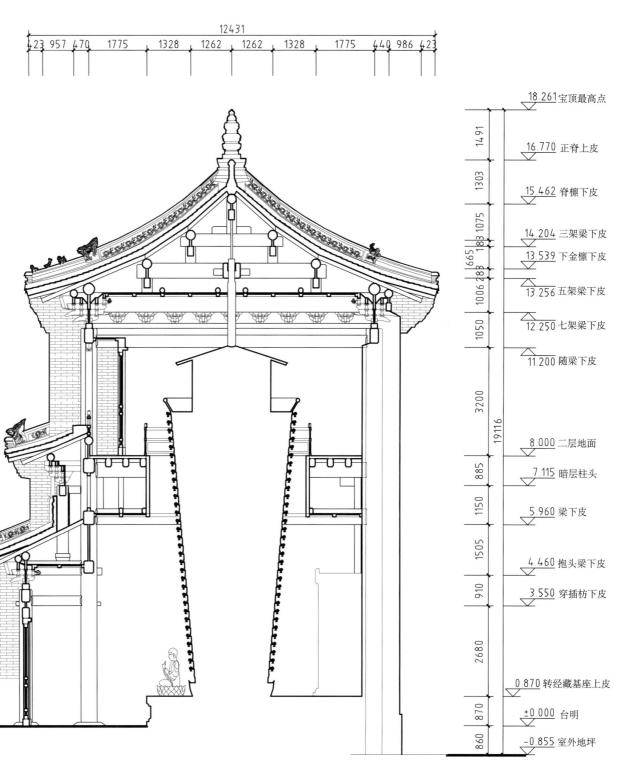

12431

| 423 | 957 | 470 | 1775 | 1328 | 1262 | 1262 | 1328 | 1775 | 440 | 986 | 423 |

宝顶最高点 18.261

14.91

正脊上皮 16.770

4.143

二层檐口 12.625

862

金柱上皮 11.765

19116

3970

夹层檐口 7.795

3190

一层檐口 4.605

565

檐柱上皮 4.040

2640

下碱 1.400

1400

台明 ±0.000

855

室外地坪 -0.855

18.261 宝顶最高点

14.91

16.770 正脊上皮

1303

15.462 脊檩下皮

1075

14.204 三架梁下皮

183

13.539 下金檩下皮

1665

13.256 五架梁下皮

288

12.250 七架梁下皮

1006

11.200 随梁下皮

1050

3200

8.000 二层地面

7.115 暗层柱头

885

5.960 梁下皮

1150

4.460 抱头梁下皮

1505

3.550 穿插枋下皮

910

2680

0.870 转经藏基座上皮

870

±0.000 台明

860

-0.855 室外地坪

19116

| 1000 | 1950 | 1761 | 6605 | 365 | 795 |

12476

塔院寺大藏经阁剖面图

Longitudinal section of *cangjingge* of Tayuansi

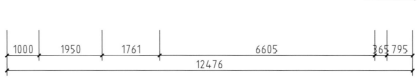

1   3m

参与测绘及相关工作的人员名单

一、测绘人员:

指导老师:刘畅 郑亮 姜东成 贺从容 包志禹 白颖 史韶华
王贵祥 李路珂 胡介中

参与学生:何潇 赵雯雯 李婷婷 李鹭飞 苏航 汪道涵 胡尚如
杨乐明 朱燕青 谢诗晨 谢宇 徐知兰 罗迪 朱勋
矫俗 邹锦铭 王智萍 李善科 徐煜坤 黄珊 邹霄
王伊倜 陈婷 李娜 戴南 蒙宇婧 万静雅 陈晓霁
明晔 梁其伟 杨松 牛牧菁 张烨 马可 包瑞
吴铭晖 谭求 谷军 方淳 陈璐 徐碧颖 邓军
周茉 郑昕 李熙 薛从余 曹楠 马获 冯纾妮
咸锈敏 祁天 卢刘颖 王逸如 睢蔚 李峰 张杨
黄逾轩 赵丽虹 张思元

二、图纸整理及相关工作

图纸统筹:李菁
图纸整理:杨博 唐恒鲁 单梦林 买琳琳 胡竞芙
英文统筹:[奥]荷雅丽
英文翻译:[奥]荷雅丽 Michael Norton

Name List of Participants Involved in Surveying and Related Works

1. Surveying and Mapping Team

Supervising Instructor: LIU Chang, ZHENG Liang, JIANG Dongcheng, HE Congrong, BAO Zhiyu, BAI Ying, SHI Shaohua, WANG Guixiang, LI Luke, HU Jiezhong

Team Members: HE Xiao, ZHAO Wenwen, LI Tingting, LI Lufei, SU Hang, WANG Daohan, HU Shangru, YANG Leming, ZHU Yanqing, XIE Shichen, XIE Yu, XU Zhilan, LUO Di, ZHU Xun, JIAO Su, ZOU Jinming, WANG Zhiping, LI Shanke, XU Yukun, HUANG Shan, ZOU Xiao, WANG Yiti, CHEN Ting, LI Na, DAI Nan, MENG Yujing, WAN Jingya, CHEN Xiaoji, MING Ye, LIANG Qiwei, YANG Song, NIU Mujing, ZHANG Ye, MA Ke, BAO Rui, WU Minghui, TAN Qiu, GU Jun, FANG Chun, CHEN Lu, XU Biying, DENG Jun, ZHOU Mo, ZHENG Xin, LI Xi, XUE Congyu, CAO Nan, MA Di, FENG Shuni, XIAN Xiumin, QI Tian, LU Liuying, WANG Yiru, JU Wei, LI Feng, ZHANG Yang, HUANG Yuxuan, ZHAO Lihong, ZHANG Siyuan

2. Editor of Drawings and Related Works

Drawings Arrangement: LI Jing
Drawings Editor: YANG Bo, TANG Henglu, SHAN Menglin, MAI Linlin, HU Jingfu
Translator in Chief: Alexandra Harrer
Translation Members: Alexandra Harrer, Michael Norton

图书在版编目（CIP）数据

五台山佛教建筑＝MOUNT WUTAI'S BUDDHIST
ARCHITECTURE/清华大学建筑学院编写；廖慧农，王贵
祥，刘畅主编．—北京：中国建筑工业出版社，2019.6
（中国古建筑测绘大系·宗教建筑）
ISBN 978-7-112-23436-3

Ⅰ.①五… Ⅱ.①清…②廖…③王…③刘… Ⅲ.
①五台山—佛教—宗教建筑—建筑艺术—图集 Ⅳ.
① TU-885

中国版本图书馆CIP数据核字（2019）第044303号

丛书策划 / 王莉慧
责任编辑 / 李 鸽 陈海娇
英文审稿 / ［奥］荷雅丽（Alexandra Harrer）
书籍设计 / 付金红
责任校对 / 王 烨

中国古建筑测绘大系·宗教建筑
五台山佛教建筑
清华大学建筑学院 编写
廖慧农 王贵祥 刘 畅 主编
Traditional Chinese Architecture Surveying and Mapping Series:
Religious Architecture
MOUNT WUTAI'S BUDDHIST ARCHITECTURE
Compiled by School of Architecture, Tsinghua University
Edited by LIAO Huinong, WANG Guixiang, LIU Chang
＊
中国建筑工业出版社出版、发行（北京海淀三里河路9号）
各地新华书店、建筑书店经销
北京方舟正佳图文设计有限公司制版
北京雅昌艺术印刷有限公司印刷
＊
开本：787毫米 ×1092毫米 横 1/8 印张：24½ 字数：649千字
2021 年 4 月第一版 2021 年 4 月第一次印刷
定价：**198.00** 元
ISBN 978-7-112-23436-3
（33732）